SpringerBriefs in Philosophy

For further volumes:
http://www.springer.com/series/10082

Anna Horská

Where is the Gödel-Point Hiding: Gentzen's Consistency Proof of 1936 and his Representation of Constructive Ordinals

 Springer

Anna Horská
Department of Logic
Charles University in Prague
Prague
Czech Republic

ISSN 2211-4548 ISSN 2211-4556 (electronic)
ISBN 978-3-319-02170-6 ISBN 978-3-319-02171-3 (eBook)
DOI 10.1007/978-3-319-02171-3
Springer Cham Heidelberg New York Dordrecht London

Library of Congress Control Number: 2013948376

© The Author(s) 2014
This work is subject to copyright. All rights are reserved by the Publisher, whether the whole or part of the material is concerned, specifically the rights of translation, reprinting, reuse of illustrations, recitation, broadcasting, reproduction on microfilms or in any other physical way, and transmission or information storage and retrieval, electronic adaptation, computer software, or by similar or dissimilar methodology now known or hereafter developed. Exempted from this legal reservation are brief excerpts in connection with reviews or scholarly analysis or material supplied specifically for the purpose of being entered and executed on a computer system, for exclusive use by the purchaser of the work. Duplication of this publication or parts thereof is permitted only under the provisions of the Copyright Law of the Publisher's location, in its current version, and permission for use must always be obtained from Springer. Permissions for use may be obtained through RightsLink at the Copyright Clearance Center. Violations are liable to prosecution under the respective Copyright Law.
The use of general descriptive names, registered names, trademarks, service marks, etc. in this publication does not imply, even in the absence of a specific statement, that such names are exempt from the relevant protective laws and regulations and therefore free for general use.
While the advice and information in this book are believed to be true and accurate at the date of publication, neither the authors nor the editors nor the publisher can accept any legal responsibility for any errors or omissions that may be made. The publisher makes no warranty, express or implied, with respect to the material contained herein.

Printed on acid-free paper

Springer is part of Springer Science+Business Media (www.springer.com)

To my mam

Acknowledgments

The work on this book was supported by the grant GA ČR 401/09/H007 Logické základy sémantiky and by Program for Development of Sciences at the Charles University in Prague no. 13 Rationality in the Human Sciences, section *Methods and Applications of Modern Logic*. Furthermore, I would like to thank Markéta Galatová, Jan von Plato, Pavel Pudlák, Vítězslav Švejdar and Jiří Velebil.

Contents

1 **Introduction** 1
   References 8

2 **Preliminaries** 11
  2.1   Sequents 11
  2.2   Initial Sequents 12
  2.3   Rules of Inference 12
  2.4   Overview of the Proof 14
  2.5   Reduction Steps for Sequents 19
  2.6   Reduction of Initial Sequents to Endform 21
  2.7   Definition of a New Derivation 24
  2.8   Chain Rule 24
     References 28

3 **Ordinal Numbers** 29
  3.1   Definition 29
  3.2   About the Ordering 31
  3.3   The Relationship Between Gentzen's Notation and Standard
       Notation of Ordinal Numbers 32
  3.4   An Algorithm for Translating Gentzen's Notation of Ordinal
       Numbers to Cantor Normal Form 35
     References 40

4 **Consistency Proof** 41
  4.1   How to Assign Ordinal Numbers to Derivations 41
  4.2   Lowering the Ordinal Numbers After Reduction Steps
       for Derivations 45

**Appendix A: Removal of Logical Operations $\vee$, $\supset$ and $\exists$
               from a Derivation** 71

Index 75

Chapter 1
Introduction

Abstract This chapter deals with brief history of Gentzen's consistency proofs and his work on proof systems which were actually byproducts of his ambition to prove the consistency of arithmetic. There is a plan in his handwritten thesis manuscript according to which he aimed to prove the consistency with the help of normalization for natural deduction. As this did not work, he developed a special semantic explanation of correctness in arithmetic. The proof based on this explanation was criticized, so Gentzen returned to an earlier idea of transfinite induction. The last proof shows directly that although the transfinite induction up to ε_0 can be formalized in arithmetic, it cannot be proved there. Further, this chapter deals with non-technical parts of his 1936 article.

Keywords Gentzen · Gentzen's thesis · Proof systems · Consistency proofs · Consistency proofs of arithmetic · History of proof systems · History of consistency proofs · Consistency of arithmetic · Hilbert's program

The trust in intuitively obvious axioms was destroyed after the discovery of paradoxes in naive set theory at the beginning of the twentieth century. Intuitions dealing with actual infinity, especially, turned out to be misleading and it was necessary to deal with them in a more careful way. On the one hand, Brouwer's *intuitionism* regards mathematics as a languageless, constructive, mental activity and intuitions of the *idealized mind* should reflect on which constructions are allowed. On the other hand, *formalism* uses axiomatic systems for mathematical investigations, but tries to eliminate all intuitions form them. This is possible in the way that all meanings are removed from an axiomatic system except for such meanings that can be defined according to the rules of the system itself. The system should be examined regardless of its interpretation. The German mathematician David Hilbert is considered to be the major proponent of formalism. He proposed a research project around 1920, today known as Hilbert's program, which was meant as an answer to the crisis brought on by the paradoxes in naive set theory. He aimed to establish secure foundations for all mathematics by grounding all theories on a finite set of axioms and schemata whose consistency should have been proved by *finitistic* methods. These methods were regarded as reliable and the consistency of the classical mathematics proved with the help of these methods would have provided a justification for non-intuitive parts of the field, especially the parts dealing with actual infinity.

In general, consistency proofs are classified into two groups: absolute and relative consistency proofs. Whereas an absolute consistency proof does not presuppose the consistency of some other system, a relative consistency proof reduces the consistency of a theory T_1 to the consistency of another theory T_2. Thus, we demand a consistency proof for T_2. In this way, we can obtain a potentially infinite hierarchy of theories. Anyway, we need a theory at the beginning of this hierarchy whose consistency can by proved without any reference to another theory, in other words, a theory which has an absolute consistency proof. It seemed that arithmetic is so fundamental that its consistency might be proved within the finitistic methods. This would suffice to regard it as an absolute consistency proof. In late 1930, Gödel's incompleteness theorems became known and it was clear that even the tools of arithmetic itself would not suffice to obtain its consistency.

Gerhard Gentzen was a student of Paul Bernays. In early 1932, he set out to work on the consistency of arithmetic. The role of Bernays in this decision of his is not known, but as Gentzen wrote in a letter to Professor Hellmuth Kneser in December 1932 (Menzler-Trott 2007, pp. 30–31), he planned to be promoted with a work on this topic by Professor Bernays. Gentzen thought this way in October 1932 (von Plato 2012, p. 315):

> The axioms of arithmetic are obviously correct, and the principles of proof obviously preserve correctness. Why cannot one simply conclude consistency, i.e., what is the true meaning of the second incompleteness theorem, the one by which the consistency of arithmetic cannot be proved by arithmetic means? Where is the Gödel-point hiding?

In contrast to his predecessors (Frege, Russell), Gentzen was not looking for absolute mathematical truth. He focused on how mathematical theorems are proved in practice, i.e., how one derives a conclusion from given assumptions. The first step he made was that he put aside prevailing proof systems of *logicism* whose linear derivations did not fit in with the idea of actual mathematical proofs because their starting points were basic logical truths and rules of inference just generated further truths. Gentzen's examination of actual reasoning ended in natural deduction calculus whose derivations were in the form of trees. He was probably inspired by Paul Hertz (1923), whose work he had been studied at the suggestion of Bernays in 1931, when he came up with the tree derivations. Although the tree derivations are in fact a departure form actual proofs, they have some technical advantages over the linear derivations: One can permute the order of rules and compose different derivations without losing a full control over the structure of the result.

The following plan is in Gentzen's handwritten thesis manuscript (von Plato 2012, p. 336):

I. To put up the calculus of natural deduction.
II. To prove that it is equivalent to standard axiomatic calculi.

III. To prove that classical arithmetic reduces to intuitionistic arithmetic.
IV. To prove normalization.
V. To extend normalization and the subformula property to arithmetic.

Let us explain the items:

I. The calculi of natural deduction, *NK* for classical and *NJ* for intuitionistic logic, were completed by September 1932. There are several possibilities which might have led Gentzen to the rules of these calculi. It is known that Gentzen studied Heyting's formalization of intuitionistic logic (Heyting 1930) whose axioms can be read as rules capturing what follows from what. Another possibility is that Gentzen converted the Brouwer-Heyting-Kolmogorov interpretation of the logical operations, which gives sufficient conditions for what does it mean to have an intuitionistic derivation of a formula, into rules. However, it is not clear whether Gentzen knew this interpretation at the time. It is also possible that Gentzen simply considered all sensible ways which were in accordance with his opinion that mathematical proofs begin with assumptions rather than axioms and 'he was a genius who got things right' (von Plato 2012, p. 322). His natural deduction calculus originally contained an induction rule. It means that Gentzen was definitely interested in arithmetic.

II. Gentzen translated derivations in natural deduction into the axiomatic logical calculus of Hilbert and Ackermann's book (Hilbert and Ackermann 1928) and back.

III. The third task was accomplished by the Gödel-Gentzen translation which was almost simultaneously developed by both mathematicians after whom it is called. The translation embeds classical arithmetic into intuitionistic one. The result is that it suffices to prove the consistency of intuitionistic arithmetic which would assure that classical arithmetic is consistent, too. We will see that Gentzen's plan failed in the item V., so he submitted this result as a separate article to the *Mathematische Annalen* in March 1933. Then he withdrew the paper because the same result by Gödel (1933) had already been published (Menzler-Trott 2007, p. 39). Gentzen's original article (Gentzen 1974b) was published by Bernays only in 1974 and in English translation in Szabo (1969).

IV. Gentzen's handwritten thesis manuscript contains a detailed proof of normalization for intuitionistic natural deduction (von Plato 2012, p. 330). Each formula in a normal derivation is a subformula of the conclusion or of an open assumption, this is called *subformula property*, and we obtain that intuitionistic logic is consistent. Since Gentzen was not able to extend the result to classical natural deduction, he developed a new sort of calculus: sequent calculus. He proved the cut elimination, his famous Hauptsatz, for intuitionistic as well as classical version of the calculus, which is a result similar to normalization. No consistency proof but his work on proof systems became the main topic of his thesis (Gentzen 1934, 1935).

It seems that these two factors inspired Gentzen to define sequents and the sequent calculus: (1) The translation from natural deduction into the axiomatic logical calculus of Hilbert and Ackermann's book (Hilbert and Ackermann 1928).

(2) Hertz-systems (Hertz 1922, 1923, 1929). As for the first aspect, the first step in the translation was to determine for each proposition A from a natural deduction derivation all those open assumptions, say $B_1..B_n$, which A depends on. Then all occurrences of A were replaced by the implication $B_1 \& .. \& B_n \supset A$ (von Plato 2012, pp. 341–342). As for the second aspect, Hertz-systems work with *sentences*, expressions similar to $B_1 \& .. \& B_n \supset A$, which allow to interpret the horseshoe as a derivability relation and the conjunction $B_1 \& .. \& B_n$ as a list of assumptions (von Plato 2012, pp. 342–343).

V. Normalization for arithmetic fails because of the rule of complete induction. This rule is actually an introduction of a formula and we have no control over the complexity of that formula. If an elimination rule follows, the induction formula need not be a subformula of the conclusion or of an open assumption and neither the subformula property nor the consistency is obtained. Gentzen probably knew this in early 1933 (von Plato 2013) and changed his intentions of writing a thesis on the consistency of arithmetic. The result was his thesis on proof systems which are accepted and generally used today.

Gentzen continued working on consistency proof of arithmetic which was finished by the end of 1934. In December 1934, he wrote a letter to Hellmuth Kneser informing him that he already had such a proof and was about to submit it to the *Mathematische Annalen* (Menzler-Trott 2007, p. 55). Gentzen's notes reveal that the very first consistency proof used the sequent calculus (von Plato 2013) instead of the natural deduction in sequent calculus style of the 1935 version which was, on the other hand, Gentzen's first submitted version. However, it remained unpublished until 1974. The original text supplemented with a prefatory words by Bernays is contained in Gentzen (1974a). Let us look at the story of the withdrawn 1935 proof.

The proof was submitted in August 1935. It was based on what Gentzen called a *reduction procedure* (Reduziervorschrift) for sequents. This procedure gives 'a special semantic explanation of correctness in arithmetic' (von Plato 2013). Sequents are expressions of the form $\Gamma \to C$, with Γ a list of assumptions and C a consequence of the assumptions. In reduction, a situation is found in which C has some false numerical consequence, say $0 = 1$. One shows that if arithmetic makes C derivable from the assumptions Γ, as expressed by $\Gamma \to C$, then Γ contains some false assumption. If Γ contains no such assumption, even C is a correct claim with no falsities hidden in it. The reduction procedure for a sequent is similar to the derivation of this sequent in Schütte's infinitary calculus of 1950 (Schütte 1951).

After Bernays, Weyl and probably Gödel and von Neumann had criticized the proof (Menzler-Trott 2007, p. 60, Footnote 150), Gentzen decided not to publish it (Menzler-Trott 2007, p. 64). The opinion of the critics was that Gentzen used the fan theorem implicitly to prove that the reduction procedure ends in a finite number of steps (Bernays 1970). The fact is that Gentzen's proof can be reformulated by the use of bar induction (von Plato 2013), a principle stronger than the fan theorem, and the fan theorem itself is too weak for Gentzen's proof.

Gentzen was not discouraged by the criticism and took up an *earlier* idea to use transfinite induction up to ε_0 to prove the termination of the reduction procedure.

1 Introduction

Indeed, first thoughts about transfinite induction can be found in Gentzen's notes already in late 1932 (von Plato 2013). The 1936 proof which uses transfinite induction appeared in Gentzen (1936) and the idea is the following. Sequents of some special kind that are derivable in arithmetic can be reduced to such simple sequents that it is easy to calculate their validity. This is possible in a finite number of steps. A sequent $\to 0 = 1$ falls within the category of these special sequents, but it is not valid and cannot be reduced. This time, the proof was accepted and 'increased Gentzen's fame' (Menzler-Trott 2007, p. 59). Through his work Gentzen made contacts with a lot of mathematicians (Menzler-Trott 2007, pp. 76–77).

Next, Gentzen tried to simplify his 1936 proof. This proof was quite difficult to follow because he changed the calculus in the middle and used an unusual notation for ordinal numbers, as he claims, to avoid *suspicious aspects of set theory* (Bedenklichkeiten der allgemeinen Mengenlehre) (Gentzen 1936, p. 555, Footnote 21). Furthermore, even if intuitionistic logic was sufficient, he used the classical natural deduction in sequent calculus style, it means no general sequents $A_1, .., A_n \to B_1, .., B_n$, and he completely avoided derivations in the form of trees. Therefore, it seems that his real intention was to write an article for a *common reader* who is familiar neither with intuitionism nor proof systems he developed in his thesis (von Plato 2012, pp. 358–359).

Thus, he published another proof in 1938. It appeared in Gentzen (1938). He made use of the classical sequent calculus *LK* and Cantor normal form. The proof is based on reductions of a putative derivation of a contradictory sequent, whereas the 1936 article contains a reduction procedure for arbitrary derivations. These reductions lead to such a simple derivation of a contradiction that it is easy to prove that it cannot exist. This version is widely known today and is treated in detail by Gaisi Takeuti (1987).

As for Gentzen's contemporaries, Ackermann (1940) and Kalmar gave alternatives to Gentzen's consistency proofs. Kalmar's proof was not published separately. It was only presented in the second volume of *Grundlagen der Mathematik* (Hilbert and Bernays 1934, 1939).

In 1939, Gentzen already wrote his article (Gentzen 1943). He showed that although the transfinite induction up to ε_0 can be formalized in arithmetic, it cannot be proved in the theory. The transfinite inductions up to any $\alpha < \varepsilon_0$, by contrast, can be formalized as well as proved in arithmetic. It yields the consistency of arithmetic because we are able to prove anything in a contradictory theory. The article was published only in 1943 and it is the beginning of ordinal analysis. Informally, the subject of ordinal analysis is to determine, for a theory T, the smallest ordinal number α such that transfinite induction up to α is sufficient to prove the consistency of T.

For further historical notes on Gentzen's proof systems, properties of his calculi, relations between them as well as Gentzen's motivations see von Plato (2012). For further Gentzen's motivations, history of early consistency proofs of arithmetic, and the role of intuitionism in the search for such proofs see von Plato (2013). For Gentzen's biography and the historical background of his work see Menzler-Trott (2007). An English translation of Gentzen's papers can be found in Szabo (1969).

The aim of this work is to provide a detailed description of the first published of Gentzen's consistency proofs. Gentzen's article which contains the proof is very comprehensive and includes philosophical as well as mathematical aspects of the problem. Gentzen divides it up into five sections.

The first section deals with reasons why consistency proofs are necessary and possible. Gentzen discusses the paradoxes in naive set theory which were the stimulus for such proofs. He stresses that a consistency proof should not only verify the axioms of the theory, but also verify the logical reasoning when applied to these axioms. The statement that a theory is consistent means that there is no derivation of a contradiction in the theory. To prove this, we must make the derivations objects of a new *meta-theory* and these derivations must be formalized properly. A consistency proof which works with the formalized derivations is just another mathematical proof that uses logical reasoning, too. It is therefore important that the reasoning in the meta-theory is more reliable than that one in the theory which is analysed. After these thoughts, Gentzen explains the consequences of Gödel's incompleteness theorems for the consistency proofs: We cannot prove the consistency of a theory purely with the tools of the theory itself.

Let us move to the second section of Gentzen's article. It treats the formalization of arithmetic. The task consists of two parts: (1) the statements of arithmetic must be formalized, (2) the process of reasoning must be formalized. Gentzen uses natural deduction in sequent calculus style mainly for expository reasons as mentioned above. He divides the inference rules into two groups: *introduction* (Einführung) and *elimination* (Beseitigung) of logical operations. Furthermore, he uses sequent calculus that displays the open assumptions in a list. By this notation, he can avoid the use of derivation trees that were prominent in his thesis (Gentzen 1934, 1935). We will describe the calculus in Chap. 2 of the present work. However, the calculus will be changed during the process of proving because of formal reasons.

The third section deals with a finitistic interpretation of the logical operations, especially the quantifiers. There is no problem when a quantifier is applied to a variable whose domain is finite. In this case, a formula $\forall x F(x)$ is a shortcut for a finite conjunction. A formula $\exists x F(x)$ stands for a finite disjunction. How should we understand the quantifiers when the domain of the bound variable is infinite? A common concept, in Gentzen's terminology *an sich-concept*, claims that a quantified formula is true or false and this is independent of whether we can decide it or not. According to this concept, a formula $\forall x F(x)$ says that every object of the infinite domain has the property F. Similarly, the formula $\exists x F(x)$ says that there exists an object with the property F in the infinite domain. This concept uses the infinity as an entity and this allows us to use the same kind of logical reasoning in the finite as well as in the infinite case. At this point, Gentzen considers the paradoxes in naive set theory and deduces that it was this concept of actual infinity that gave rise to them. Therefore, he rejects this concept and suggests using the concept of potential infinity. Based on this concept, he proposes a finitistic interpretation of the logical operations. He combines semantic explanations (for $\forall$, $\exists$, &, $\vee$) with an explanation in terms of provability (for $\supset$, $\neg$), today known as Brouwer-Heyting-Kolmogorov interpretation.

1 Introduction

A formula $\forall x F(x)$ has the following finitistic meaning: When we start with 0 and substitute this one and the following numbers for x in $F(x)$ stepwise, we always obtain a true statement. A formula $\exists x F(x)$ means that we have found a number n with the property F although n is not mentioned explicitly in the quantified formula. There is no difficulty with the finitistic interpretation of conjunction $A \& B$: This one is valid whenever A, B are both valid. A finitistic interpretation of disjunction is similar to the one for $\exists$: A formula $A \vee B$ is valid whenever A or B has already been recognized as valid.

The interpretations of implication and negation are not so straightforward. A formula of the form $A \supset B$ seems to say that if we have a proof of A, we are able to transform it into a proof of B. However, the proof of B may also include some rules for $\supset$ or A might be of the form $C \supset D$. This can make the interpretation circular. A formula of the form $\neg A$ means that the assumption A leads to a contradiction. This corresponds with the interpretation of the formula $A \supset 0 = 1$. So, all difficulties with implication are carried over into the interpretation of negation. Gentzen, in fact, had a proof of normalization for intuitionistic natural deduction by which the explanation is not circular in the case of pure logic. He further argues that the rules of the calculus (defined in Chap. 2 of the present work) are in harmony with the finitistic interpretation of the logical operations. Nevertheless, there is one exception. The rule *Beseitigung der doppelten Verneinung* allows us to conclude A from $\neg\neg A$. It is not clear why this should hold because there may not be a direct proof of A. According to Gentzen, the task of the consistency proof is to verify logical reasoning including the steps which we do not have a proper finitistic interpretation for, i.e., the application of double negation elimination to transfinite propositions and the use of transfinite propositions containing nested operations $\supset$ and $\neg$.

The fourth section of Gentzen's article contains the consistency proof itself. A detailed analysis of the proof is the main part of the present work.

In the fifth section, Gentzen comments on the steps of his proof. He particularly focuses on these two aspects: (1) To what extent are the steps reliable? (2) How do they exceed the theorems of arithmetic in connection with Gödel's theorem? The part that needs to be examined most is the proof that the reduction procedure ends in a finite number of steps. In order to prove this, Gentzen used transfinite induction up to ε_0 which does not seem to be a finitistic method. Gentzen argues that the ordinal numbers are built constructively: A number β is *accessible* (erreichbar) only when we already know that all $\alpha < \beta$ are accessible. To show that a number α is accessible, we have to pass through all numbers less than α. Limit ordinal numbers are also reached because we can follow the whole infinite sequence before the limit ordinal arbitrarily far. So, we are allowed to regard every member of the sequence as accessible. Gentzen believes that this concept is transparent enough to be considered reliable. As for the second aspect, Gentzen claims that the proof is in accordance with Gödel's theorem because there is no obvious way how the transfinite induction up to ε_0 can be proved in arithmetic in contrast to other principles and steps used. Indeed, Gentzen showed directly that it is impossible to prove it there three years later. He ends his article with thoughts about the possibilities of how the method of the proof can be useful to fields of mathematics other than arithmetic and discusses objections of intuitionism to such proofs.

As far as Gentzen's 1936 proof is concerned, the idea and the results are preserved in this paper. However, Gentzen omitted quite a few proofs and provided just informal explanations. We turned his informal sketches into lemmas, especially Lemma e1, Lemma e2, and Lemma e6–e7 (see Sect. 4.2). The key property of reduction steps is that they lower ordinal numbers of derivations. The proof of this fact needed revising in two unclear cases (e3 and e4) and we added exact calculations (Lemma e3, Lemma e4) where Gentzen had omitted them. There is an interesting relationship between Gentzen's representation of ordinal numbers and the set-theoretical representation which was claimed by Gentzen, but he did not provide any proof. The proof is in Sect. 3.3. Furthermore, we defined an explicit algorithm for translating Gentzen's representation into Cantor normal form (see Sect. 3.4). This algorithm provides a more detailed elaboration of the relationship between the two different representations of ordinal numbers.

References

Ackermann, W. 1940. Zur Widerspruchsfreiheit der Zahlentheorie. *Mathematische Annalen* 117: 162–194.
Bernays, P. 1970. On the original Gentzen consistency proof for number theory. In *Intuitionism and proof theory*, ed. J. Myhill, 409–417. Amsterdam: North-Holland.
Gentzen, G. 1934. Untersuchungen über das logische Schliessen I. *Mathematische Zeitschrift* 39: 176–210.
Gentzen, G. 1935. Untersuchungen über das logische Schliessen II. *Mathematische Zeitschrift* 39: 405–431.
Gentzen, G. 1936. Die Widerspruchsfreiheit der reinen Zahlentheorie. *Mathematische Annalen* 112: 493–565.
Gentzen, G. 1938. Neue Fassung des Widerspruchsfreiheitsbeweises für die reine Zahlentheorie. *Forschungen zur Logik und zur Grundlegung der exakten Wissenschaften* 4: 19–44.
Gentzen, G. 1943. Beweisbarkeit und Unbeweisbarkeit von Anfangsfällen der transfiniten Induktion in der reinen Zahlentheorie. *Mathematische Annalen* 119: 140–161.
Gentzen, G. 1974a. Der erste Widerspruchsfreiheitsbeweis für die klassische Zahlentheorie. *Archiv für Mathematische Logik* 16: 97–118.
Gentzen, G. 1974b. Über das Verhältnis zwischen intuitionistischer und klassischer Arithmetik. *Archiv für Mathematische Logik* 16: 119–132.
Gödel, K. 1933. Zur intuitionistischen Arithmetik und Zahlentheorie. *Ergebnisse eines mathematischen Kolloquiums* 4: 34–38.
Hertz, P. 1922. Über Axiomensysteme für beliebige Satzsysteme I, Sätze ersten Grades. *Mathematische Annalen* 87: 246–269.
Hertz, P. 1923. Über Axiomensysteme für beliebige Satzsysteme II, Sätze höheren Grades. *Mathematische Annalen* 89: 76–102.
Hertz, P. 1929. Über Axiomensysteme für beliebige Satzsysteme. *Mathematische Annalen* 101: 457–514.
Heyting, A. 1930. Die formalen Regeln der intuitionistischen Logik. In *Sitzungsberichte der Preussischen Akademie der Wissenschaften*, 42–65.
Hilbert, D., W. Ackermann. 1928. *Grundzüge der theoretischen Logik*. Berlin: Springer.
Hilbert, D., P. Bernays. 1934. *Grundlagen der Mathematik I*. Berlin: Springer.

References

Hilbert, D., P. Bernays. 1939. *Grundlagen der Mathematik II*. Berlin: Springer.
Menzler-Trott, E. 2007. *Logic's lost genius: The life of Gerhard Gentzen*. Providence: American Mathematical Society.
Schütte, K. 1951. Beweistheoretische Erfassung der unendlichen Induktion in der Zahlentheorie. *Mathematische Annalen* 122: 369–389.
Szabo, M.E. 1969. *The collected papers of Gerhard Gentzen*. Amsterdam: North-Holland.
Takeuti, G. 1987. *Proof theory*. Amsterdam: North-Holland.
von Plato, J. 2012. Gentzen's proof systems: Byproducts in a work of genius. *Bulletin of Symbolic Logic* 18: 313–367.
von Plato, J. 2014. From Hauptsatz to Hilfssatz. In *The Quest for consistency*, eds. M. Baaz, R. Kahle, M. Rathjen, Springer, to appear.

Chapter 2
Preliminaries

Abstract The calculus *NLK* and further notions preliminary to the 1936 consistency proof are defined in this chapter. The most important notion is *endform* which represents sequents whose validity can be decided. Reduction steps for sequents, whose task is to reduce sequents to endform, are presented. Furthermore, an algorithm for reducing initial sequents to endform is defined and a detailed overview of the consistency proof is given. The chapter ends with a modification of the calculus and the most important rule of the new calculus, *chain rule*, which can be seen as a generalized cut, is discussed. The modification of the calculus is necessary because it makes the definition of reduction steps for derivations easier.

Keywords Sequent · Calculus *NLK* · Natural deduction · Natural deduction in sequent calculus style · Endform · Chain rule · Reduction steps · Reduction steps for sequents

2.1 Sequents

We start by defining a natural deduction in sequent calculus style for Peano arithmetic (PA) denoted by *NLK* in Gentzen's handwritten thesis manuscript. The language of PA is $L = \{+, \cdot, S, =, 0\}$, where $+$ and $\cdot$ are binary functional symbols, S is an unary functional symbol, $=$ is a binary relational symbol and 0 is a constant symbol. Every sequent (Sequenz) has to appear in the following form

$$A_1, .., A_n \rightarrow B$$

where $A_1, .., A_n$ are antecedent formulas (Vorderformeln). We view them as a list. It is possible that they are missing. The formula denoted by B is a succedent formula (Hinterformel). There must always be exactly one succedent formula in a sequent.

The calculus *NLK* includes two kinds of initial sequents: logical and mathematical. It also includes three kinds of inference rules: structural (Strukturänderungen), logical and an induction rule.

2.2 Initial Sequents

Logical initial sequents are of the form $D \to D$, where D is an arbitrary formula in L. Mathematical initial sequents are of the form $\to C$, where C is a mathematical axiom. Gentzen does not introduce a particular list of mathematical axioms. He claims that it is not important for the proof which set of axioms we choose (Gentzen 1936, p. 519). We shall choose equality axioms and Robinson arithmetic axioms to obtain a common axiomatization of PA. The schema of induction is represented by a special rule.

Equality axioms:

- $\forall x (x = x)$
- $\forall x \forall y (x = y \supset y = x)$
- $\forall x \forall y \forall z (x = y \,\&\, y = z \supset x = z)$
- $\forall x_1 .. \forall x_n \forall y_1 .. \forall y_n (x_1 = y_1 \,\&\, .. \,\&\, x_n = y_n \supset F(x_1, .., x_n) = F(y_1, .., y_n))$
  where F is a n-ary functional symbol in L.
- $\forall x_1 .. \forall x_n \forall y_1 .. \forall y_n (x_1 = y_1 \,\&\, .. \,\&\, x_n = y_n \,\&\, R(x_1, .., x_n) \supset R(y_1, .., y_n))$
  where R is a n-ary relational symbol in L.

Robinson arithmetic axioms:

- $\forall x \forall y (S(x) = S(y) \supset x = y)$
- $\forall x \neg (S(x) = 0)$
- $\forall x (\neg x = 0 \supset \exists y (S(y) = x))$
- $\forall x (x + 0 = x)$
- $\forall x \forall y (x + S(y) = S(x + y))$
- $\forall x (x \cdot 0 = 0)$
- $\forall x \forall y (x \cdot S(y) = x \cdot y + x)$

2.3 Rules of Inference

Although Gentzen does not use tree derivations in Gentzen (1936), not even inference lines, we shall stick to this notation as it is common today.

Definition 2.1 Upper sequents in each rule of inference are called *premises*. The lower sequent is called *conclusion*.

Structural rules (Strukturänderungen) of *NLK* are given in Table 2.1.

2.3 Rules of Inference

Table 2.1 Structural rules of *NLK*

Exchange (Vertauschen)	$\dfrac{\Gamma, A, C, \Delta \to B}{\Gamma, C, A, \Delta \to B}\ Ex$
Contraction (Weglassen)	$\dfrac{\Gamma, A, A, \Delta \to B}{\Gamma, A, \Delta \to B}\ Ct$
Weakening (Zufügen)	$\dfrac{\Gamma \to B}{\Gamma, A \to B}\ Wk$
Renaming of bound variables	–

The German names in Table 2.1 are intuitive explanations used by Gentzen (1936). He used these words to describe the effect of the rules because he has no formal notation for derivations in the article. Gentzen's standard terminology elsewhere is *Vertauschung*, *Zusammenziehung* and *Verdünnung*, respectively.

Induction rule:

$$\frac{\Gamma \to F(0) \quad F(a), \Delta \to F(a+1)}{\Gamma, \Delta \to F(t)}$$

Here a is an eigenvariable and must, therefore, not occur in Γ, Δ, $F(0)$, $F(t)$, and t is an arbitrary term in L that can be substituted for a in $F(a)$.

Table 2.2 Logical rules of *NLK*

	Introduction	Elimination
&	$\dfrac{\Gamma \to A \quad \Theta \to B}{\Gamma, \Theta \to A\&B}\ \&I$	$\dfrac{\Gamma \to A\&B}{\Gamma \to A}\ \&E$
∨	$\dfrac{\Gamma \to A}{\Gamma \to A \vee B}\ \vee I$	$\dfrac{\Gamma \to A \vee B \quad A, \Delta \to C \quad B, \Theta \to C}{\Gamma, \Delta, \Theta \to C}\ \vee E$
⊃	$\dfrac{A, \Gamma \to B}{\Gamma \to A \supset B}\ \supset I$	$\dfrac{\Gamma \to A \quad \Delta \to A \supset B}{\Gamma, \Delta \to B}\ \supset E$
¬	$\dfrac{A, \Gamma \to B \quad A, \Delta \to \neg B}{\Gamma, \Delta \to \neg A}\ \neg I$ [a]	$\dfrac{\Gamma \to \neg\neg A}{\Gamma \to A}\ \neg E$ [b]
∀	$\dfrac{\Gamma \to F(a)}{\Gamma \to \forall x F(x)}\ \forall I$ [c]	$\dfrac{\Gamma \to \forall x F(x)}{\Gamma \to F(t)}\ \forall E$ [d]
∃	$\dfrac{\Gamma \to F(t)}{\Gamma \to \exists x F(x)}\ \exists I$ [d]	$\dfrac{\Gamma \to \exists x F(x) \quad F(a), \Delta \to C}{\Gamma, \Delta \to C}\ \exists E$ [c]

[a] Gentzen calls this rule *Widerlegung*
[b] Gentzen calls this rule *Beseitigung der doppelten Verneinung*
[c] The variable a is an eigenvariable and must, therefore, not occur in $\Gamma, \forall x F(x)$ and $\Gamma, \Delta, C, \exists x F(x)$, respectively
[d] The symbol t stands for a term in L that can be substituted for x in F

Logical rules of *NLK* are given in Table 2.2. The rule of cut in sequent calculus is: $\frac{\Gamma \to A \quad A, \Sigma \to C}{\Gamma, \Sigma \to C}$. It is a derivable rule in Gentzen's calculus: $\frac{\Gamma \to A \quad \frac{A, \Sigma \to C}{\Sigma \to A \supset C}}{\Gamma, \Sigma \to C}$.

Definition 2.2 *A derivation* (Herleitung) in calculus *NLK* is a treelike structure that consists of sequents. Each sequent is an initial sequent or is derived from previous ones using one of the rules of inference. The last sequent of a derivation is called *endsequent* (Endsequenz).

We simplify our calculus by removing the logical symbols $\vee$, $\supset$ and $\exists$. This is easy: We use relations between the logical operations valid in classical predicate logic. All occurrences of the rules for the defined operations will be substituted by short pieces of correct derivations that do not contain any of the defined logical operations. A formal execution of this transformation can be found at the end of this book (Appendix A). Thus, we assume that the allowed rules of inference mentioned in Definition 2.2 are only those for the operations &, $\neg$ and $\forall$.

Definition 2.3 A sequent is said to be in *endform* (Endform) when the following conditions are met: It does not contain any free variables. Its succedent formula is a true equation, or its succedent formula is a false equation and there is at least one false equation among the antecedent formulas.

It is obvious that a sequent in endform is a valid sequent because it contains a false formula in the antecedent or a true formula in the succedent. Whether an atomic sentence, an equation without free variables in this case, is true or false can be easily calculated because the functional symbols are represented as primitive recursive functions.

Definition 2.4 Let A be an arbitrary formula in L. The expression $|A|$ stands for the number of logical operations in A.

Note the difference between the notion of *derivation* and the notion of *proof*. A derivation is a formal treelike structure in the calculus (Definition 2.2). The proof is the whole consistency proof at a meta-level outside of PA.

2.4 Overview of the Proof

In this section, we give a rather informal overview of Gentzen's proof. Although we have defined the calculus *NLK*, the actual proof does not work with derivations in this calculus. The calculus will be changed (Sects. 2.7, 2.8) and it actually means that we can forget about *NLK* after the modification is done, because the whole reduction process is defined using derivations in the new calculus. Gentzen did not explain the reasons for the change in his article (Gentzen 1936). It is highly probable that, instead of writing a new paper, he rewrote parts of his 1935 proof after it had been criticized. This resulted in the 1936 article. The modification of the 1935 paper was possible because the notion of reduction steps for sequents, which also played a role in the

2.4 Overview of the Proof

1935 proof, is independent of the calculus (von Plato 2013). We decided to preserve the midway modification of the calculus in the present work because (1) we want to stay as close to the original article as possible and (2) even if we introduced only the new calculus, we would have to prove its equivalence with first-order predicate logic. Thus, we would not avoid defining any of the standard calculi. Let us call a derivation according to Definition 2.2 an *old derivation*. Similarly, a derivation in the new calculus is a *new derivation*.

The main difference between the two calculi, namely *NLK* and the new one, is that almost all logical rules from *NLK* disappear and are replaced by *groundsequents* (Sect. 2.7) in order to make the new calculus. The most important inference rule in the new calculus is a *chain rule* by which many of the rules from Sect. 2.3 are derivable. Of course, the chain rule cooperates with the added groundsequents while simulating the original rules. The chain rule can be seen as a generalized cut. As expected, we are able to translate every old derivation into a new derivation.

So, we know that reduction steps are defined on new derivations, but this is not the whole thing. We have to apply a further small modification to new derivations: We shall replace all free variables except eigenvariables by arbitrary numerals and calculate the value of any term whenever it is possible.[1] In the end, the terms will be replaced by numerals with the corresponding values. So, we substitute general propositions by their numerical instances. There are universal quantifiers and eigenvariables left in the derivation. In regard to this matter, the reduction steps will be helpful. We hope to obtain formulas without variables eventually, because it is easy to decide whether they are true or false. The later analysis of reduction steps is based on the notion of new derivation after this replacement and the calculation.

Note that if the induction rule had been used in a new derivation, its use could have failed after the replacement of terms. Assume that a formula $F(a)$ with an eigenvariable a contains a term $a + x$ that has the form $a + \bar{4}$ after the replacement. A formula $F(t)$ has to appear in the conclusion of the induction rule. The term t is an arbitrary term in L and the eigenvariable a was substituted by it. Let t be $\bar{2} \cdot \bar{3}$, so we have $\bar{2} \cdot \bar{3} + \bar{4}$. We go on to calculate the value and obtain $\overline{10}$. It means that a numeral $\overline{10}$ stands at the position where $t + x$ was before and the formula $F(t)$ is not in its previous form anymore. In the following text, we allow to use the induction rule in this way.

The proof will show that every sequent that can be obtained by a new derivation after the replacement and the calculation can be reduced to endform in a finite number of steps. A sequent in endform will not be reduced anymore. This form represents the simplest form of a sequent and its characteristic feature is that we can easily see that the sequent is valid.

It suffices to consider new derivations after the replacement and the calculation because a derivation of a contradiction, represented by the sequent $\to 0 = 1$, would also be of this form. If there existed an old derivation of $\to 0 = 1$, we would be able to transform it into a new derivation. The calculation and the replacement do not change the endsequent $\to 0 = 1$ because $0 = 1$ is an atomic sentence and other

---

[1] It is impossible when there is a bound variable or an eigenvariable in the term.

formulas, with or without free variables, disappear before the end of the derivation. The point is that $\rightarrow 0 = 1$ is not in endform and it is impossible to reduce it either. We will see it immediately after the reduction steps are defined.

Above, we gave a sketch of what we would like to prove. Let us describe the method of the proof in more detail now. We said that we want to reduce the endsequent of a derivation until we reach endform. (1) We shall introduce *reduction steps for sequents* which will not be defined uniquely. It means that we can choose which reduced form of the sequent is helpful for us in every special case and sometimes, we have more than one acceptable possibility. Gentzen calls this property *Wahlfreiheit*, i.e., freedom of choice. A disadvantage is that we are not always able to find the right form immediately and must, therefore, maintain the possibility of a new choice. (2) We also show a method for modifying derivations: *reduction steps for derivations*. These steps transform a derivation to another correct derivation such that its endsequent remains the same or will be changed according to exactly one reduction step for sequents.

Gentzen defined an algorithm for reducing the initial sequents to endform (Sect. 2.6). Since an initial sequent is a derivation, this tells us what reduction steps for such simple derivations look like. Every reduction step carried out on an initial sequent leads to another initial sequent. To modify more complex derivations, we use the induction hypothesis: *It is possible to carry out a reduction step for derivations if the endsequent of the derivation we are about to reduce is not in endform*. There are, roughly speaking, three kinds of reduction steps for derivations. All of them examine the last inference rule used in the derivation and, based on this rule, they decide what to do:

- The first possibility is that we immediately know which reduction step for sequents applies to the endsequent. Then, this reduction step for sequents gives us the reduction for the whole derivation: It consists in omitting the endsequent. Could we discard the contradiction in this way if it was the endsequent? No, it cannot be the case, because this step is used only when the endsequent contains at least one logical operation before we apply this modification to the derivation.
- The second possibility is that the last inference rule is the induction rule. Then its conclusion has a numerical term $\bar{n}$ and the induction can be replaced by a series of cuts. Since we want to get rid of the induction rule and do not want to replace it with n separated cuts, we use the *chain rule*. It allows us to simulate n cuts at once. So, a derivation whose last inference rule is an induction rule will be reduced to a derivation whose last inference rule is a chain rule. The endsequent remains unchanged.
- The third possibility is that we are not able to reduce the endsequent straightaway. In this case, there exists a premise of the endsequent that is not in endform, and therefore, by the induction hypothesis, we are able to modify its derivation and reduce the premise. There are several possible forms that the reduced premise can have. We have to consider all of these and show how the reduced premise can be integrated into the last inference rule the endsequent is derived by. The endsequent of the reduced derivation does not need to be the same as in the original derivation,

2.4 Overview of the Proof

but it must have one of the acceptable forms represented by the reduction steps for sequents.

Let us look closer at the reduction steps for sequents (Sect. 2.5). They are applicable to sequents without free variables and without the operations $\exists$, $\vee$ and $\supset$. In general, their task is to simplify sequents: In most cases, a sequent contains fewer logical operations after the reduction step. We introduce two kinds of reduction steps for sequents. We shall use Gentzen's original numbering for them: 13.2 and 13.5.

Steps 13.2 examine the succedent formula of a sequent. If it is not an atomic sentence, then it has one of these forms: $\forall x F(x)$, $A_1 \& A_2$, $\neg A$. The step prescribes to replace it by $F(\bar{n})$, A_1 or A_2, $0 = 1$ (plus we have to add A among the antecedent formulas), respectively. Note that the reduction is valid for an arbitrary choice of $\bar{n}$ and A_i. This is the mentioned freedom of choice and it plays a role in the reductions for derivations (Case e3 in Sect. 4.2). It does not mean that we can put there whatever we want whenever we want. However, if other circumstances of the reduction process force us to substitute for example $\bar{3}$ for x in $F(x)$, then it is a valid reduction step. Note that steps 13.2 have the ability to turn an invalid sequent into a valid one. Since they modify only sequents whose succedent formula contains a logical operation, we do not need to worry that a possible falsity $\to 0 = 1$ will be deleted.

The second group of reduction steps for sequents is numbered 13.5. These steps are applicable to the sequents whose succedent formula is a false atomic sentence. If it was a true atomic sentence, the whole sequent would be in endform and we would not need to reduce it anymore. It is clear, therefore, that steps 13.5 examine the antecedent formulas. Similarly to 13.2, they look for a formula of the form $\forall x F(x)$, $A_1 \& A_2$, $\neg A$ in the antecedent. Then, exactly one formula from the antecedent is chosen: $\forall x F(x)$ is replaced by $F(\bar{n})$, $A_1 \& A_2$ is replaced by A_1 or A_2, and $\neg A$ changes the succedent into A. So, these reduction steps proceed similar to steps 13.2. Nevertheless, there are two crucial differences:

1. We have no freedom of choice. The instance to choose is prescribed by circumstances of the reduction process and only this possibility counts as a valid reduction step. On the one hand, instances in steps 13.5 are often determined by choices in steps 13.2. On the other hand, determined instances in steps 13.5 affect the choices in steps 13.2. This is not a circle as we will proceed by induction on the derivation and, roughly speaking, reduction steps for premises of a rule of inference will affect reduction step for the conclusion of the rule.
2. In some delicate cases, a copy of the reduced formula stays among the antecedent formulas (Case e2 in Sect. 4.2). Unfortunately, the sequent will not contain fewer logical operations after the reduction, but we can do without.

Note that there is no reduction step for $\to 0 = 1$.

Why do we have to allow such reduction steps for sequents that do not lower the amount of logical operations? We intuitively see that steps 13.2 produce a valid sequent under the assumption that the sequent to reduce is valid.[2] If $\forall x F(x)$ holds, then each instance $F(\bar{n})$ holds. Therefore, if $\Gamma \to \forall x F(x)$ is valid, so is $\Gamma \to F(\bar{n})$.

---

[2] A valid sequent has a false formula in the antecedent or a true formula in the succedent.

We would like to maintain this property in steps 13.5, too. However, these steps are not as crystal clear as 13.2. The succedent formula is a false atomic sentence, so, there must be a false formula hidden in the antecedent of a valid sequent. To preserve the validity of the sequent after the reduction step, we must not delete the false antecedent formula. If the false formula has the form $\forall x\, F(x)$, then we must choose an instance that is false again. If we do not know this instance, we must keep the whole formula $\forall x\, F(x)$ so that we can revise our choice later.

The initial sequents that are reduced according to 13.5 never make use of letting the reduced formula stay in the antecedent. We are always able to find the false instance of the false formula. The only case where we have to apply this is the case e2: a reduction of the endsequent that was derived by the chain rule. It keeps the chain rule correct after the reduction step.

Using the notion of reduction steps for sequents, we shall define reduction steps for derivations (Sect. 4.2). Recall: These steps change a derivation to another correct derivation and its endsequent remains the same or will be changed according to exactly one reduction step for sequents. The entities that are factually reduced all the time are only sequents and nothing else. However, reducing the sequents in the context of the whole derivation helps us to assign ordinal numbers to the individual stages of the reduction in a sensible way. So, we will be able to prove the termination of the process.

If we combine the way the reduction steps for derivations use the reduction steps for sequents with the effect of the reduction steps for sequents, then it follows that the goal of the reductions is to change the sequents from the derivation to such sequents that contain only atomic sentences. Atomic sentences are not changed anymore and their validity can be decided.

The last segment of the proof is to show that the reduction process is finite. For the purpose of demonstrating this, we use the ordinal numbers less than ε_0. Gentzen invented a notation for ordinal numbers based on decimal numbers syntactically (Sect. 3.1). Each derivation is assigned an ordinal number (Sect. 4.1). This number represents the complexity of the derivation. The reduction steps for derivations simplify the complexity of derivations: They lower their ordinal numbers (Sect. 4.2). The ordinal numbers are well-ordered (Sect. 3.3). So, in a finite number of steps, we will obtain a derivation that is not reducible anymore. It is clear that the endsequent of this reduced derivation is in endform. If it were not, we would be able to reduce it again and this would, of course, decrease its ordinal number.

Let us summarize the purpose of the construction: We have a derivation in PA and this derivation has an endsequent. We have to consider all allowed rules that the endsequent could have been derived by and define reduction steps for derivations according to the form of the endsequent. We do this under the assumption that the endsequent is not in endform. It is possible that the endsequent is a falsity, like $\rightarrow 0 = 1$. The aim is to show that it cannot be the case, but we do not know this at the beginning and we must, therefore, take the possibility into account. There are several possible rules that a falsity could be derived by. The reduction steps for derivations whose last inference rule is one of these are defined so that the endsequent remains unchanged. So, we would be able to reduce the derivation of the

2.4 Overview of the Proof

contradiction repeatedly and this would construct an infinite decreasing sequence of ordinal numbers which is impossible.

2.5 Reduction Steps for Sequents

We use Gentzen's original numbering 13.2 and 13.5 for referring to reduction steps for sequents. Gentzen marked with a number almost every paragraph in his article (Gentzen 1936) and he did it very meticulously. Paragraphs 1–11 discuss finitistic methods and the necessity for consistency proofs. Paragraph 12 deals with the removal of the logical operations $\vee$, $\supset$ and $\exists$ from a derivation. That is why the numbers of the paragraphs that contain definitions of reduction steps for sequents are relatively high. Gentzen's section 13.4 defines the notion of endform.

Reduction steps for sequents without free variables are defined in the following way:

- 13.21 If the succedent formula is of the form $\forall x F(x)$, it is replaced by the formula $F(\bar{n})$ where $\bar{n}$ is an arbitrary numeral:

$$\Gamma \to \forall x F(x) \quad \leadsto \quad \Gamma \to F(\bar{n})$$

- 13.22 If the succedent formula is of the form $A \& B$, it is replaced by the formula A or B. The reduction is valid for both alternatives:

$$\Gamma \to A \& B \quad \leadsto \quad \Gamma \to A \text{ or } \Gamma \to B$$

- 13.23 If the succedent formula is of the form $\neg A$, it is changed to a false atomic sentence and the formula A is added to the antecedent formulas:

$$\Gamma \to \neg A \quad \leadsto \quad \Gamma, A \to 0 = 1$$

After these reductions, the sequent has fewer logical operations. If no case mentioned above is applicable, the succedent formula of the examined sequent must be an atomic sentence. If it is a true atomic sentence, the sequent is already in endform and there is no need to define a reduction step for such a sequent. Assume it is a false atomic sentence. We go on to define further reduction steps. This time, they are based on examining the antecedent formulas.

- 13.51 If there is a formula of the form $\forall x F(x)$ among the antecedent formulas, the sequent acquires one of these forms:

$$\Gamma, \forall x F(x) \to 0 = 1 \quad \leadsto \quad \begin{cases} \Gamma, \forall x F(x), F(\bar{n}) \to 0 = 1 \\ \\ \Gamma, F(\bar{n}) \to 0 = 1 \end{cases}$$

It will be specified, for each concrete case, whether the formula $\forall x\, F(x)$ is deleted and what numeral $\bar{n}$ is substituted for x.

- 13.52 If there is a formula of the form $A \& B$ among the antecedent formulas, the sequent acquires one of these forms:

$$\Gamma, A\&B \to 0 = 1 \quad \leadsto \quad \begin{cases} \Gamma, A\&B, A \to 0 = 1 \\ \Gamma, A \to 0 = 1 \\ \Gamma, A\&B, B \to 0 = 1 \\ \Gamma, B \to 0 = 1 \end{cases}$$

It will be specified, for each concrete case, which possibility should occur. An alternative reduction would be:

$$\Gamma, A\&B \to 0 = 1 \quad \leadsto \quad \Gamma, A, B \to 0 = 1$$

However, we stick to Gentzen's definition because of the symmetric treatment of cases 13.51 and 13.52.

- 13.53 If there is a formula of the form $\neg A$ among the antecedent formulas, the sequent acquires one of these forms:

$$\Gamma, \neg A \to 0 = 1 \quad \leadsto \quad \begin{cases} \Gamma, \neg A \to A \\ \Gamma \to A \end{cases}$$

It will be specified, for each concrete case, which possibility should occur.

Definition 2.5 The formula used in these reduction steps for sequents is *affected formula*.

Thus, the affected formulas in cases 13.51, 13.52 and 13.53 are $\forall x\, F(x)$, $A\&B$ and $\neg A$, respectively.

We shall attempt to reduce sequents in such a way that the affected formula disappears. However, sometimes we will not be able to avoid keeping it among the antecedent formulas. If a formula that already stands among the antecedent formulas is created by the reduction, we will not record it again.

These are all the reduction steps for sequents. Note that $\to 0 = 1$ cannot be reduced and is not in endform.

Let us denote the cases 13.21, 13.22 and 13.23 all at once by the shortcut 13.2, and similarly for 13.5.

2.6 Reduction of Initial Sequents to Endform

We define an algorithm for reducing all initial sequents to endform. This algorithm will be used for defining reduction steps for derivations.

Lemma 2.1 *Every initial sequent without free variables can be reduced to endform in a finite number of steps.*

Proof Let us begin with the logical initial sequents. Consider $D \rightarrow D$, where D is a formula that does not contain any free variables. We start by executing steps 13.2 and will carry on until the succedent formula D turns into $\neg C$ or into an atomic sentence. Then, three cases can occur:

1. If our succedent formula is a true atomic sentence, we have endform.
2. If our succedent formula is a false atomic sentence, the sequent is of the form: $D \rightarrow m = n$ with $m = n$ false. This can be replaced by $0 = 1$. We go on to reduce the antecedent formula D: Steps 13.5 are executed. The antecedent formula D has to be modified in the same way as the succedent formula D was at the beginning of the process. It means that the same changes have to be applied to the antecedent formula in the same order as while modifying the succedent formula D according to steps 13.2. Of course, the same choices have to be made. This leads to endform because the antecedent formula D becomes a false atomic sentence, too.
3. Assume our succedent formula is $\neg C$. The sequent is of the form: $D \rightarrow \neg C$. We execute step 13.23 and obtain $D, C \rightarrow 0 = 1$. At this point, we apply steps 13.5 under the same conditions as in the second case described: The antecedent formula D turns into $\neg C$ because the succedent formula D in the original sequent $D \rightarrow D$ turned into $\neg C$ after we had executed steps 13.2. Now, we have the sequent $\neg C, C \rightarrow 0 = 1$. After executing step 13.53, we obtain $C \rightarrow C$. This is a logical initial sequent. Note that the formula C contains at least one logical operation less than the formula D. The reason for this is that $\neg C$ resulted from the modification of D:

$$|D| \geq |\neg C| > |C|.$$

The whole process has to be repeated with the sequent $C \rightarrow C$. After a finite number of steps, we reach endform.

Let us look at the mathematical initial sequents $\rightarrow C$, where C is an equality axiom or a Robinson arithmetic axiom. In general, we proceed the way described below. The only exception is the initial sequent $\rightarrow \forall x(\neg x = 0 \supset \exists y(S(y) = x))$. It is reduced in Examples 2.2 and 2.3. The reduction proceeds as follows:

- Write the axiom without the logical operations $\vee, \supset$ and $\exists$. This is a preparation step.
- All variables are bound by the universal quantifiers and all axioms are in prenex normal form. It is clear that the first reduction steps will be 13.21: We substitute numerals for all variables. These steps lead to sequents without any variables. Free

variables were not there anyway and a removal of all bound variables is what we just did.
- We go on to execute steps 13.2 until it is impossible. Note that if we apply step 13.2 to a valid sequent, we obtain a valid sequent again. It is evident that all mathematical initial sequents are valid.
- At some point, when it becomes necessary to use step 13.5 for the first time, we have a valid sequent whose succedent formula is a false atomic sentence. It means that there is at least one false antecedent formula in the sequent. We are able to find out which one is false because the whole sequent does not contain any variables. This false antecedent formula will be used as the affected formula (Definition 2.5) in the following reduction step. If there is a choice to make, we make it in the way that guarantee the falsity of the formula created by the reduction step.
- The formula affected by the reduction is always deleted and never stays among the antecedent formulas. □

Example 2.1 Let us reduce the sequent $\to \forall x \forall y (S(x) = S(y) \supset x = y)$.

- $\to \forall x \forall y \neg(S(x) = S(y) \,\&\, \neg x = y)$ *Preparation, removal of the operation $\supset$.*
- $\to \neg(S(0) = S(0) \,\&\, \neg 0 = 0)$ *steps 13.21*
- $S(0) = S(0) \,\&\, \neg 0 = 0 \to 0 = 1$ *step 13.23*
- $\neg 0 = 0 \to 0 = 1$ *step 13.52*
- $\to 0 = 0$ *step 13.53*

Example 2.2 Let us reduce the sequent $\to \forall x(\neg x = 0 \supset \exists y(S(y) = x))$.

- $\to \forall x \neg(\neg x = 0 \,\&\, \forall y \neg(S(y) = x))$ *Preparation, removal of the operation $\supset$.*
- $\to \neg(\neg 0 = 0 \,\&\, \forall y \neg(S(y) = 0))$ *step 13.21*
- $\neg 0 = 0 \,\&\, \forall y \neg(S(y) = 0) \to 0 = 1$ *step 13.23*
- $\neg 0 = 0 \to 0 = 1$ *step 13.52*
- $\to 0 = 0$ *step 13.53*

Example 2.3 Let us give another reduction for the sequent $\to \forall x(\neg x = 0 \supset \exists y(S(y) = x))$.

- $\to \forall x \neg(\neg x = 0 \,\&\, \forall y \neg(S(y) = x))$ *Preparation, removal of the operation $\supset$.*
- $\to \neg(\neg \bar{5} = 0 \,\&\, \forall y \neg(S(y) = \bar{5}))$ *step 13.21*
- $\neg \bar{5} = 0 \,\&\, \forall y \neg(S(y) = \bar{5}) \to 0 = 1$ *step 13.23*
- $\forall y \neg(S(y) = \bar{5}) \to 0 = 1$ *step 13.52*
- $\neg(S(\bar{4}) = \bar{5}) \to 0 = 1$ *step 13.51*
- $\to S(\bar{4}) = \bar{5}$ *step 13.53*

Note that we have to reach endform no matter what we have chosen in steps 13.21 and 13.22. Compare Examples 2.2 and 2.3.

In the remaining part of this section, we show how sequents $A \& B \to A$; $\forall x F(x) \to F(\bar{n})$; $A, B \to (A \& B)$; $A, \neg A \to 0 = 1$; $\neg\neg A \to A$ can be reduced to endform. We need this because they will be used later, some of them

2.6 Reduction of Initial Sequents to Endform

modified a little. Let us call them *groundsequents*. Note that they are all derivable in NLK.

Lemma 2.2 *Let the groundsequents $A\&B \to A$; $\forall x F(x) \to F(\bar{n})$; $A, B \to (A\&B)$; $A, \neg A \to 0 = 1$; $\neg\neg A \to A$ be without free variables. Then they can be reduced to endform in a finite number of steps.*

Proof Let us analyse the groundsequents individually:

1. $A\&B \to A$; $\forall x F(x) \to F(\bar{n})$: These groundsequents occur whenever steps 13.22 and 13.21 are applied to the logical initial sequents $A\&B \to A\&B$ and $\forall x F(x) \to \forall x F(x)$, respectively. After that, the reduction method is the same as if we continued reducing the mentioned initial sequents.

2. $A, B \to (A\&B)$: First, we apply step 13.22 to this groundsequent. The result is $A, B \to A$ or $A, B \to B$ depending on our choice. Second, we continue reducing as if we had one of these logical initial sequents: $A \to A$; $B \to B$. There is a redundant antecedent formula in our sequent, but it does not matter. It is not used while reducing.

3. $A, \neg A \to 0 = 1$: After applying step 13.53, we simply reduce the logical initial sequent $A \to A$.

4. $\neg\neg A \to A$: We apply steps 13.2 until the succedent formula A turns into an atomic sentence or has the form $\neg C$. We have to examine three possibilities:

- If the succedent formula is a true atomic sentence, we reached endform.
- If the succedent formula is a false atomic sentence, we obtained the sequent $\neg\neg A \to 0 = 1$. We use step 13.53 and have $\to \neg A$. After that, we use step 13.23. It leads to $A \to 0 = 1$. We know that the formula A changed to a false atomic sentence after steps 13.2 had been applied. It implies that reducing the initial sequent $A \to A$ according to the algorithm described in the proof of Lemma 2.1 would lead to the sequent $A \to 0 = 1$, and this is exactly the sequent that we currently have. So, we go on to reduce it as if we had started with $A \to A$.
- Assume finally that the succedent formula turned into $\neg C$. We obtained $\neg\neg A \to \neg C$. The series of reduction steps: 13.23 $\langle \neg\neg A, C \to 0 = 1 \rangle$, 13.53 $\langle C \to \neg A \rangle$, 13.23 $\langle A, C \to 0 = 1 \rangle$ leads to the sequents in brackets, respectively. Note that formula A acquired the form $\neg C$ after reduction steps 13.2 had been applied. And this is the point again. We can imagine that we had reduced the initial sequent $A \to A$ all the time. Since we obtained $A, C \to 0 = 1$, it is now sufficient to follow the steps of the algorithm for reducing the logical initial sequents. □

We know how to reduce an initial sequent and a groundsequent to endform in a finite number of steps. Especially important is that the formula affected by the reduction step always disappears. This ensures that the number of the logical operations in the reduced sequent decreases after each step.

2.7 Definition of a New Derivation

We are going to change the allowed initial sequents and the rules of inference for the sake of simplifying the definition of reduction steps for derivations. In this way, we obtain the notion of *new derivation*. Furthermore, we show that each old derivation can be transformed into a new derivation in such a way that both have identical endsequents. This is needed because we do not want to lose any sequent derivable in PA.

New derivation is a treelike structure that consists of sequents. Every sequent is a *new initial sequent* or is derived from previous sequents by a *new rule of inference*.

The new mathematical initial sequents are all old mathematical initial sequents, after the logical operations $\vee$, $\supset$ and $\exists$ were removed from them, and all sequents that result from applying to them a finite number of reduction steps for sequents.

New logical initial sequents are, first, all old logical initial sequents and groundsequents:

- $D \rightarrow D$
- $A \& B \rightarrow A$
- $A \& B \rightarrow B$
- $A, B \rightarrow A \& B$
- $\forall x F(x) \rightarrow F(t)$
- $A, \neg A \rightarrow 0 = 1$
- $\neg\neg A \rightarrow A$

Second, we take all sequents that result from applying a finite number of reduction steps for sequents to the list above.

We take just two old inference rules, namely $\forall$-introduction and the induction rule, among the new rules. Furthermore, we define two new rules: a *chain rule* (Kettenschluß) and a *negation rule*. The negation rule is of the form:

$$\frac{\Gamma, A \rightarrow 0 = 1}{\Gamma \rightarrow \neg A}$$

So, we have four rules allowed in a new derivation: $\forall$-introduction, the induction rule, the negation rule and the chain rule.

2.8 Chain Rule

Informally, the chain rule is a generalized cut. Let us explain how it works before we give a formal definition:

Let a sequence of sequents $\Gamma_1 \rightarrow A_1 .. \Gamma_n \rightarrow A_n$ be given, with $n \geq 1$. This sequence represents the premises of the chain rule. Let us describe how the chain rule derives a conclusion from them. Pick out one sequent from the premises, assume it is $\Gamma_i \rightarrow A_i$, for some $i \leq n$. This sequent is called the *main premise*. The succedent formula of the conclusion is the same as the succedent formula of the main premise.

2.8 Chain Rule

If A_i is a false atomic sentence, we can use an arbitrary false atomic sentence instead of A_i. The antecedent formulas of the conclusion are created as follows: Take all formulas from $\Gamma_1 .. \Gamma_i$ and arrange them arbitrarily. Each formula has to be marked, so that we know which sequence Γ_j it comes from. For example, a formula B_j originates from the sequence Γ_j. Now, we go through the sequence of the arranged formulas. Assume the formula B_j is in turn. If B_j is already among the antecedent formulas of the conclusion, we do not put it there again. If B_j is the succedent formula of any sequent indexed by $k < j$, we do not put it there either. In other cases, B_j has to be written among the antecedent formulas of the conclusion. Finally, we are allowed to put arbitrary formulas among the antecedent formulas of the conclusion and rename bound variables. The chain rule is given by the following scheme:

$$\frac{\Gamma_1 \to A_1 \;..\; \boxed{\Gamma_i \to A_i} \;..\; \Gamma_n \to A_n}{\Delta \to A_i}$$

The main premise is placed in a frame to make the use of the chain rule clearer. The premises that stand on the right of the main premise are not used, but we cannot delete them as the ordinal numbers of their derivations are used while reducing the whole derivation.

A formal definition follows:

Definition 2.6 *Chain rule* is of the form:

$$\frac{\Gamma_1 \to A_1 \;..\; \boxed{\Gamma_i \to A_i} \;..\; \Gamma_n \to A_n}{\Delta \to A_i}$$

Here $i \leq n$ and all formulas in Γ_k are among $\Delta, A_1, .., A_{k-1}$ for $k = 1, .., i$. The premise $\Gamma_i \to A_i$ is called the *main premise*.

Example 2.4

$$\frac{A, B \to D \quad A \to B \quad \boxed{D, E \to C} \quad D \to E}{A, B, E \to C}$$

Explanation:

1. The sequent $D, E \to C$ is the main premise.
2. The premise $D \to E$ has no effect because it stands on the right of the main premise.
3. The sequent $A, B, E \to C$ is the conclusion. Note that its succedent formula is the same as in the main premise, namely C.
4. Formulas relevant to the antecedent of the conclusion are: A, B, A, D, E. They are the antecedent formulas of the premises: $A, B \to D$; $A \to B$; $D, E \to C$. Assume we have ordered them the way written earlier. It is necessary to examine which of them we should include in the antecedent of the conclusion.

5. The first is A. It is neither in the antecedent of the conclusion nor in the succedent of any premise situated on the left of $A, B \to D$, because there is no premise on the left of the first one. We include A in the conclusion.
6. The second is B. The same case as the previous one.
7. The third is A. There is already a formula A among the antecedent formulas of the conclusion. So, we do not duplicate its occurrence.
8. The fourth is D. A formula D is the succedent formula of the premise $A, B \to D$ situated on the left of the sequent where our D comes from. That is why we do not put D among the antecedent formulas of the conclusion.
9. The fifth is E. This is a case similar to item 5. Note that to make a decision whether to put E among the antecedent formulas of the conclusion, we must examine the succedent formulas of the premises on the left of the main premise, because E comes from the main premise.
10. We did not use the option of adding additional formulas among the antecedent of the conclusion.

Lemma 2.3 *Let a sequence of sequents $\Gamma_1 \to A_1 .. \Gamma_m \to A_m$ be given, with $m \geq 1$. We apply the chain rule twice using this sequence as the premises. First, the main premise is $\Gamma_n \to A_n$, for some $n \leq m$ and we obtain $\Gamma \to A_n$. For the second time, the main premise is $\Gamma_i \to A_i$, for some $i < n$. We obtain $\Gamma^\star \to A_i$. If no additional formulas are put among the antecedent formulas of the conclusions, this is generally allowed by the chain rule, then all formulas in $\Gamma^\star$ are contained in Γ.*

Proof The first application of the chain rule has the form:

$$\frac{\Gamma_1 \to A_1 .. \Gamma_i \to A_i .. \boxed{\Gamma_n \to A_n}, \Gamma_{n+1} \to A_{n+1} ..}{\Gamma \to A_n}$$

The second application of the chain rule has the form:

$$\frac{\Gamma_1 \to A_1 .. \boxed{\Gamma_i \to A_i} .. \Gamma_n \to A_n, \Gamma_{n+1} \to A_{n+1} ..}{\Gamma^\star \to A_i}$$

We use indirect proof. Assume there is a formula B such that $B \in \Gamma^\star$ and $B \notin \Gamma$. Then we have $B \in \Gamma_1$ or .. or $B \in \Gamma_i$, as $B \in \Gamma^\star$. However, $B \notin \Gamma$ and this implies that B must have been deleted in the first chain rule. This is possible only when B occurs as the succedent formula in some sequent that stands on the left of the premise that contains B among its antecedent formulas. This would cause that B would also have been deleted in the second chain rule and hence would not be in $\Gamma^\star$. This contradicts our assumption $B \in \Gamma^\star$. □

We would like to show now that every old derivation can be translated into a new derivation.

Lemma 2.4 *Each old derivation can be transformed into a new derivation in such a way that the endsequents are the same in both.*

2.8 Chain Rule

Proof Let us take an old derivation. The logical and the mathematical initial sequents in the old derivation are correct initial sequents for a new derivation. There are, of course, other rules in our old derivation that are not allowed to occur in a new derivation. They will therefore be replaced, especially by the chain rule. We begin with the structural rules:

- *Exchange*: $\frac{\Gamma, A, B, \Delta \to C}{\Gamma, B, A, \Delta \to C}$, the main premise of the chain rule is the only premise of the structural rule. The chain rule allows us to arrange the formulas in the antecedent of the conclusion arbitrarily. So, we choose the way that creates the right conclusion.
- *Contraction*: $\frac{\Gamma, A, A, \Delta \to B}{\Gamma, A, \Delta \to B}$, the main premise of the chain rule is the only premise of the structural rule. The chain rule prescribes not to double the antecedent formulas of the conclusion. Thus, we must leave out the second occurrence of formula A.
- *Weakening*: $\frac{\Gamma \to B}{\Gamma, A \to B}$ is changed into

$$\frac{A \to A \quad \boxed{\Gamma \to B}}{\Gamma, A \to B}$$

- *Renaming of bound variables*: $\frac{\Gamma, \forall x F(x), \Delta \to A}{\Gamma, \forall y F(y), \Delta \to A}$, the main premise of the chain rule is the only premise of the structural rule. The antecedent formulas of the conclusion are ordered in the same way as the antecedent formulas of the main premise. The chain rule allows us to rename the bound variable x.

There are five logical rules that are not allowed in a new derivation, but they can occur in an old derivation. These rules are: &-introduction, &-elimination, ∀-elimination, ¬-introduction, ¬-elimination. In order to remove them, we replace them by the chain rule:

- &-introduction: $\frac{\Gamma \to A \quad \Theta \to B}{\Gamma, \Theta \to A \& B} \quad \leadsto \quad \frac{\Gamma \to A \quad \Theta \to B \quad \boxed{A, B \to A \& B}}{\Gamma, \Theta \to A \& B}$

- &-elimination: $\frac{\Gamma \to A \& B}{\Gamma \to A} \quad \leadsto \quad \frac{\Gamma \to A \& B \quad \boxed{A \& B \to A}}{\Gamma \to A}$

- ∀-elimination: $\frac{\Gamma \to \forall x F(x)}{\Gamma \to F(t)} \quad \leadsto \quad \frac{\Gamma \to \forall x F(x) \quad \boxed{\forall x F(x) \to F(t)}}{\Gamma \to F(t)}$

- ¬-introduction: $\frac{A, \Gamma \to B \quad A, \Delta \to \neg B}{\Gamma, \Delta \to \neg A} \quad \leadsto \quad \frac{A, \Gamma \to B \quad A, \Delta \to \neg B \quad \boxed{B, \neg B \to 0=1}}{\frac{\Gamma, \Delta, A \to 0=1}{\Gamma, \Delta \to \neg A}}$

- ¬-elimination: $\frac{\Gamma \to \neg \neg A}{\Gamma \to A} \quad \leadsto \quad \frac{\Gamma \to \neg \neg A \quad \boxed{\neg \neg A \to A}}{\Gamma \to A}$

The main premises of the chain rules above are groundsequents. We used the negation rule in the last step while simulating the ¬-introduction. □

In the following chapters, we are going to work with new derivations.

References

Gentzen, G. 1936. Die Widerspruchsfreiheit der reinen Zahlentheorie. *Mathematische Annalen* 112: 493–565.
von Plato, J. 2014. From Hauptsatz to Hilfssatz. In *The Quest for consistency*, ed. M. Baaz, R. Kahle, M. Rathjen, Springer, to appear.

Chapter 3
Ordinal Numbers

Abstract Gentzen's representation of ordinal numbers less than ε_0, which is based on decimal numbers syntactically, is given. Systems σ which provide information about the ordering of the numbers in Gentzen's notation are introduced. The relationship between Gentzen's representation and Cantor normal form is analysed with the help of systems σ and a recursive algorithm for translating Gentzen's notation to Cantor normal form is defined. Furthermore, correctness of this algorithm is proved and some easy examples of how it works are presented.

Keywords Ordinals · Constructive ordinals · Ordinal numbers · Ordinal numbers less than ε_0 · Gentzen's representation of ordinal numbers · Cantor normal form · Well-ordering · Transfinite induction

3.1 Definition

We shall demonstrate that repeated reductions of a derivation lead to a derivation whose endsequent is in endform. Therefore, we need to show that every reduction step for derivations makes the derivation simpler in some way. Every derivation gets an ordinal number assigned which represents how complex the derivation is. Reduction steps for derivations will be defined to lower the ordinal numbers of derivations. Ordinal numbers are well-ordered and do not build an infinite decreasing sequence. So, in a finite number of steps, a derivation will be reached that we are unable to reduce. The endsequent of this derivation is in endform. If it were not, there would be a reduction step available that would decrease the ordinal number again.

A definition of Gentzen's notation of ordinal numbers less than ε_0 follows. Let us call them just ordinal numbers.

3 Ordinal Numbers

Definition 3.1 *Ordinal numbers* are some of the positive decimal numbers (gewisse positive endliche Dezimalbrüche). The part that stands on the left of the decimal mark is called a *numerus* (Numerus).[1] The part that stands on the right of the decimal mark is called a *mantissa* (Mantisse):

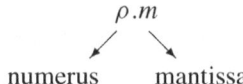

Assume m_1 and m_2 are mantissae. If for the decimal numbers $0.m_1$ and $0.m_2$ it holds that $0.m_1 < 0.m_2$, we say that the mantissa m_1 is smaller than the mantissa m_2: $m_1 < m_2$.

- Ordinal numbers with the numerus 0 are: 0.1, 0.11, 0.111 .. 0.2. These are all numbers whose mantissa contains only a finite number of digits 1. The number 0.2 is the only number with the numerus 0 that contains a digit other than 1 in its mantissa.
- Assume that the numbers with the numerus ρ are defined. Take an arbitrary finite number $k \geq 1$ of ordinal numbers with the numerus ρ that are all different from each other and order their mantissae according to their values: from the greatest to the smallest. Assume their ordered mantissae are: $m_1 > m_2 > .. > m_k$.

A new ordinal number with the numerus $\rho + 1$ is created this way:

$$\rho + 1 . m_1 \underbrace{0..0}_{\substack{\text{Write exactly } \rho + 1 \\ \text{0's between each pair} \\ \text{of the mantissae.}}} m_2 \, 0 ..0 \, m_3 \, ..m_k$$

We do not add the digits 0 to the end of any mantissa. Although its value would not change, we require that it is defined unambiguously. If we have a number with numerus $\rho + 1$, it is easy to find out which numbers with numerus ρ it was made from.

Note that only the digits 0, 1 and 2 are allowed to be in a mantissa. Furthermore, the digits 1 and 2 are never immediate neighbours. It holds true because the mantissa of each number with the numerus 0 consists purely of 1's or of the solitary digit 2. The other ordinal numbers are composed of the mantissae of some numbers with a smaller numerus and these mantissae are separated by sequences of 0's.

Ordinal numbers are ordered in the same way as we would order them if they were decimal numbers. We assume the standard order of decimal numbers.

Example 3.1

- The number 1.1102 *does not* represent an ordinal number. It is built of the ordinal numbers 0.11 and 0.2, but their mantissae are not ordered.

---

[1] Latin: number.

3.1 Definition   31

- The number 1.202 *does not* represent an ordinal number. It is built of the ordinal number 0.2, but this number was used twice. It is not allowed. All mantissae that we create a new number from must be different.
- The number 1.20011 *does not* represent an ordinal number. It is built of the ordinal numbers 0.2 and 0.11. Their mantissae are ordered. However, the mantissa 20011 contains this wrong sequence 00. A sequence that consists of 0's cannot be longer than the value of the numerus. Note that the maximal sequence of 0's can be shorter than the value of the numerus. It occurs when the new ordinal number is built of only a *single* ordinal number with a smaller numerus. This is clear: We had just one mantissa so we did not need to use separating sequences of 0's. For instance, let us look at the number 2.201. It is correct and was built of 1.201. For creating 1.201, we used 0.2 and 0.1.
- The number 3.20111101002001110100011 *does* represent an ordinal number. It is built of the numbers 2.201111010020011101 and 2.11. The first one is built of 1.20111101, 1.2 and 1.11101. It is clear that the second one is built of 1.11.

3.2 About the Ordering

Assume $\alpha = (\rho.m)$ is an ordinal number with a numerus $\rho \geq 0$ and with a mantissa m. Let us define a system $\sigma(\alpha)$ that belongs to α:

$$\alpha \rightsquigarrow \sigma(\alpha)$$

Here $\sigma(\alpha)$ is a system that contains such ordinal numbers with the numerus $\rho+1$ that have the following property: *The greatest* mantissa used for creating them is exactly the same as the ordinal number α has. Assume there is a number β in the system $\sigma(\alpha)$. So, if we reformulate it a little, it means that the ordinal number α is one of the ordinal numbers with the numerus ρ which were used for building β and the mantissa of α is *the greatest* among the mantissae which participated in building β:

$$\sigma(\alpha) = \{(\rho + 1.\bar{m}) \mid \bar{m} = m \underbrace{0..0}_{\rho+1} ..\}$$

We emphasize that the mantissa $\bar{m}$ begins with the mantissa m, the mantissa of α, and the rest can be arbitrary (provided that a correct ordinal number will be created). Every ordinal number with the numerus $\rho + 1$ belongs to such a system and this system is given unambiguously. The idea of systems σ comes from Gentzen.

Lemma 3.1 *Let α_1 and α_2 be ordinal numbers. Then:*

$$\alpha_1 < \alpha_2 \text{ implies } \forall \beta_1 \in \sigma(\alpha_1) \, \forall \beta_2 \in \sigma(\alpha_2) \, (\beta_1 < \beta_2).$$

Proof When we consider the assumption $\alpha_1 < \alpha_2$, two possibilities can occur:

- The numbers α_1 and α_2 have the same numerus ρ and the mantissa of α_1 is smaller than the mantissa of α_2. It follows that both β_1 and β_2 have the same numerus $\rho + 1$. The mantissa of β_1 begins with the mantissa of α_1 and the mantissa of β_2 begins with the mantissa of α_2. Since the mantissa of α_1 is smaller than the mantissa of α_2, we obtain $\beta_1 < \beta_2$.
- The number α_1 has a smaller numerus than α_2. It is obvious that the numerus of β_1 is smaller than the numerus of β_2 and that is why β_1 is smaller than β_2. □

It follows that the ordering of the ordinal numbers can be extended to an ordering of the systems σ in a very nice and natural way.

Let us investigate the ordering of the numbers within a single system $\sigma(\alpha)$. The smallest number in this system is $\alpha + 1$. The other numbers are built of $\alpha + 1$ by adding to the end of the mantissa a sequence of 0's that is $\rho + 1$ in length. Furthermore, we have to add the mantissa of an ordinal number with the numerus $\rho + 1$ that is smaller than $\alpha + 1$ to this sequence of 0's:

$$\alpha = (\rho.m)$$
$$\beta = (\rho + 1.\bar{m}), \beta < (\alpha + 1)$$

the smallest in $\sigma(\alpha)$
$\rightsquigarrow$

$$\alpha + 1 = (\rho + 1.m)$$
$$\gamma = \rho + 1.m \underbrace{0\,..\,0}_{\rho+1}\bar{m},\ \gamma \in \sigma(\alpha)$$

Note that there is a connection between the way the ordinal numbers $< \alpha + 1$ with the numerus $\rho + 1$ are ordered and the ordering of the members in the system $\sigma(\alpha)$.

3.3 The Relationship Between Gentzen's Notation and Standard Notation of Ordinal Numbers

We show that Gentzen's representation covers exactly ordinal numbers less than ε_0.

Definition 3.2 ε_0 is the supremum of the sequence $\omega, \omega^\omega, \omega^{(\omega^\omega)}, \omega^{(\omega^{(\omega^\omega)})} \ldots$

Definition 3.3 We say that two ordered sets a and b have the same *order type* whenever they are isomorphic, i.e., there exists a bijection $f : a \to b$ such that both f and f^{-1} are order-preserving.

For a discussion of ordinal numbers and ordinal arithmetic see Jech (2003).

The following relationship was mentioned as a footnote in Gentzen (1936) on p. 555. However, Gentzen did not provide any proof.

Theorem 3.1 *The ordinal numbers with the numerus ρ are of an order type 2^α where α is the order type of the ordinal numbers with the numerus $\rho - 1$:*

3.3 The Relationship Between Gentzen's Notation and Standard Notation

numerus 0 $\rightsquigarrow$ *type* $\omega + 1$

numerus 1 $\rightsquigarrow$ *type* $2^{\omega+1} = \omega + \omega$

numerus 2 $\rightsquigarrow$ *type* $2^{\omega+\omega} = \omega \cdot \omega$

numerus 3 $\rightsquigarrow$ *type* $2^{\omega \cdot \omega} = \omega^{\omega}$

numerus 4 $\rightsquigarrow$ *type* $2^{(\omega^\omega)} = \omega^{(\omega^\omega)}$

$\vdots$

Proof The ordinal numbers with the numerus 0 have the order type $\omega + 1$ because of the following isomorphism:

$$\begin{array}{ccccccc} 0.1 & 0.11 & 0.111 & 0.1111 & & 0.2 \\ \downarrow & \downarrow & \downarrow & \downarrow & \cdots & \downarrow \\ 0 & 1 & 2 & 3 & & \omega \end{array}$$

This ordering corresponds to the order type $\omega + 1$ because $\omega + 1$ as a set contains all natural numbers and their limit ω, too.

Let us deal with the ordinal numbers with the numerus 1. As said, their order type is $\omega + \omega$. To prove this, we are going to construct them with the help of the systems σ. Each system is numbered by its index:

0. $\sigma(0.1) = \{1.1\}$
1. $\sigma(0.11) = \{1.11, 1.1101\}$
2. $\sigma(0.111) = \{1.111, 1.11101, 1.111011, 1.11101101\}$
3. $\sigma(0.1111) = \{1.1111, 1.111101, 1.1111011, 1.111101101, 1.11110111,$
   $1.1111011101, 1.11110111011, 1.1111011101101\}$

$\vdots$

ω. $\sigma(0.2) = \{1.2, 1.201, 1.2011, 1.201101, 1.20111, 1.2011101, 1.20111011,$
   $1.2011101101, 1.201111, 1.20111101, ..\}$

We prove by induction on the index of the system that every system except $\sigma(0.2)$ has just a finite number of elements, namely 2^n, where n is the index of the given system.

We have obviously: $|\sigma(0.1)| = 1 = 2^0$. Assume that the systems with the indices $j < i < \omega$ contain 2^j elements, respectively. Let us calculate how many elements there are in the system with the index i. The smallest number in this system is fresh unlike the other numbers that are based on the numbers from the previous systems: Every number from every previous system gave birth to exactly one number from the examined system with the index i. The system with the index i has therefore $1 + 2^0 + 2^1 + 2^2 + .. + 2^{i-1} = 1 + \frac{2^i - 1}{2 - 1} \cdot 1 = 1 + 2^i - 1 = 2^i$ elements.

If we take all numbers from these finite systems and order them as if they were decimal numbers, we see that the order type of this union is ω. The way the system $\sigma(0.2)$ is created reveals that its order type is $1 + \omega = \omega$.

It follows that the order type of the ordinal numbers with the numerus 1 is $\omega + \omega$:

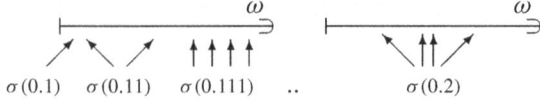

We have: $2^{\omega+1} = 2^\omega \cdot 2 = \sup\{2^n; n < \omega\} \cdot 2 = \omega \cdot 2 = \omega + \omega$. Since $\omega + 1$ is the order type of the numbers with the numerus 0 and $\omega + \omega$ is the order type of the numbers with the numerus 1, we obtained our desired result.

Let us continue with the induction argument. Assume that the ordinal numbers with the numerus ρ have an order type α. We wish to show that the ordinal numbers with the numerus $\rho+1$ have the order type 2^α. The ordinal numbers with the numerus $\rho + 1$ can be created from the ordinal numbers with the numerus ρ as a union of the systems σ:

$$\sigma(\rho.m_0) \; \sigma(\rho.m_1) \; \sigma(\rho.m_2) \; .. \sigma(\rho.m_n) \; .. \sigma(\rho.m_\omega) \; ..$$

These systems are ordered as the ordinal numbers with the numerus ρ. They are of the order type α. Similar to the finite systems, we want to show that a system with the index β has the order type 2^β. We already have it for the indices 0, 1, .. ω. Let us take $\beta > \omega$ and assume that it holds for every system whose index is smaller than β. The number β is either a successor ordinal or a limit ordinal.

- Assume that β is a successor ordinal. Then a number γ exists such that $\beta = \gamma + 1$. The induction hypothesis tells us that the order type of the system with the index γ is 2^γ. Note: The initial segment of the system with the index β is of the order type 2^γ too, because its origin is based on elements in the same previous systems as those which were used for building the whole system with the index γ. The mentioned initial segment is followed by elements that are built of the numbers contained in the system with the index γ itself. Finally, we obtain that the system $\sigma(\rho.m_\beta)$ is of the order type $2^\gamma + 2^\gamma = 2^\gamma \cdot 2 = 2^{\gamma+1} = 2^\beta$.
- Assume that β is a limit ordinal. The induction hypothesis tells us that the order type of any system with the index $\gamma < \beta$ is 2^γ. The order type of the system with the index β is the same as if we put all elements from the systems with the indices $\gamma < \beta$ in a row. Actually, each system is constructed in this way. It means that the order type of the system with the index β is the supremum of the ordinals that represent the order types of all systems with an index below β: $\sup\{2^\gamma; \gamma < \beta\} = 2^\beta$. The equality is apparent from the definition of exponentiation.

Let us create now an imaginary system. Its index will be α and it will contain the ordered elements from the systems σ which belong to the numbers with the numerus ρ and whose indices are $\gamma < \alpha$. It means that it will contain all ordinal numbers with the numerus $\rho + 1$. Note: This imaginary system is created in the same way as

3.3 The Relationship Between Gentzen's Notation and Standard Notation   35

every system that precedes it. Therefore, we know that the order type of this system must be 2^α where α is its index. However, α is the order type of the numbers with the numerus ρ. This is our desired result. □

3.4 An Algorithm for Translating Gentzen's Notation of Ordinal Numbers to Cantor Normal Form

We shall present an algorithm that translates every ordinal number written in Gentzen's notation to Cantor normal form.

Definition 3.4 The *Cantor normal form* of an ordinal number $\beta > 0$ is

$$\beta = \sum_{i=1}^{n} \omega^{\delta_i} \cdot v_i$$

where $n, v_i \neq 0; n, v_i \in \omega; \delta_1 > \delta_2 > .. > \delta_n$.

Every system of ordinal numbers with a particular numerus ρ has its smallest number $\rho.1$. Based on the previous section, we know which Cantor normal form belongs to these smallest numbers:

- 0.1 has 0
- 1.1 has $\omega + 1$
- 2.1 has $\omega \cdot 3$
- 3.1 has $\omega \cdot \omega$
- 4.1 has ω^ω
- 5.1 has $\omega^{(\omega^\omega)}$
- ..

Now, we have an estimate for intervals which we want to map the systems with different numeri onto. Let us introduce the steps of the *algorithm*:

- We have a number $\rho.m$ whose Cantor normal form is to be determined. Find out the numbers $(\rho - 1.m_1) > .. > (\rho - 1.m_n)$ that the number $\rho.m$ is composed of. Determine the numbers $a_1 > .. > a_n$ which indicate the positions of $(\rho-1.m_1)..(\rho-1.m_n)$ in the system of the numbers with the numerus $\rho - 1$. It means that $a_1..a_n$ are not their global positions in the ordering of all ordinal numbers. They are relevant just to the system of the numbers with the particular numerus $\rho - 1$.
- If a_i is finite, define $b_i = 2^{a_i} - 1$. Otherwise, $b_i = 2^{a_i}$.
- The Cantor normal form, which we are looking for, is the result of the following ordinal addition:

$$(\omega + 1) + (\omega + \omega) + (\omega \cdot \omega) + \ldots + \underbrace{\gamma}_{\substack{\text{ordering of the} \\ \text{numbers with the} \\ \text{numerus } \rho - 1}} + (b_1 + 1) + (b_2 + 1) + \ldots + b_n$$

Note that this is a recursive algorithm. To determine the positions $a_1..a_n$, we have to calculate the Cantor normal forms $\beta_1..\beta_n$ which correspond to $(\rho-1.m_1)..(\rho-1.m_n)$. This is possible because we reach numbers with the numerus 0 in the worst case during the calculation and we know their positions in their corresponding system. Since we are familiar with the number δ that corresponds to $(\rho-1).1$, we are able to find the only one a_i such that

$$\delta + a_i = \beta_i$$

What is the intuitive idea behind this algorithm? First, we sum up the order types of all systems that have a smaller numerus than the number $\rho.m$ we examine. Second, we have to focus on the system with the numerus ρ itself and investigate the line we jumped in order to get directly to $\rho.m$. It is helpful to consider the way new numbers are generated. Let us look at this easy example:

The number 1.1111011101 is made from $\alpha_1 = 0.1111, \alpha_2 = 0.111$ and $\alpha_3 = 0.1$. If we consider the system with the numerus 0, we find out that the position of α_1 in this particular system is 3. It means that there are three smaller mantissae available for creating ordinal numbers with the numerus 1. Numbers built of these smaller mantissae will be smaller than 1.1111011101 for sure. Three smaller mantissae we are talking about are 111, 11, 1 and new numbers with the numerus 1 smaller than 1.1111011101 are created this way:

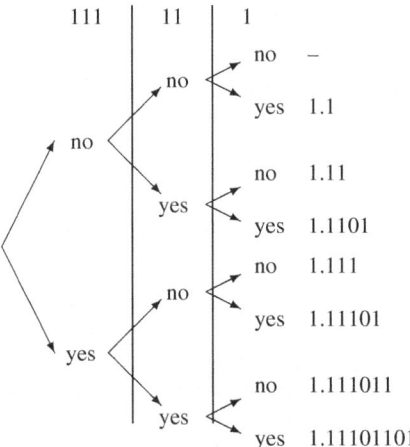

Their order type is $2^3 - 1 = 7$ because the item 'no, no, no' is not a mantissa. The next possibility we have to take into account is the number 1.1111. This is the reason why there is the expression '+1' in the algorithm. We continue the same way: The

3.4 An Algorithm for Translating Gentzen's Notation of Ordinal Numbers

second mantissa to examine is that of $\alpha_2 = 0.111$. Its position in the system with the numerus 0 is 2:

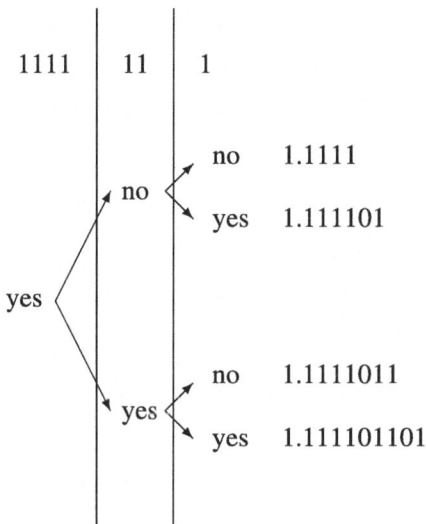

Since the item 'yes, no, no' was taken into account in the last step, their order type is $2^2 - 1 = 3$. We must not forget about the possibility 1.11110111 itself. When we try to examine the mantissa of $\alpha_3 = 0.1$, we discover that there are no smaller mantissae than this one in the system with the numerus 0. The calculation matches this fact: $2^0 - 1$. No '+1' occurs behind this last expression because this would stand for the whole examined mantissa 1111011101 itself and we are interested only in mantissae that are smaller.

The '−1' followed by the '+1' looks unnatural, but we want to keep it, to maintain the similarity with the notation for the infinite case.

To recapitulate the idea demonstrated with the help of this example, we can say that numbers with the same numerus as the examined number that are smaller than this one are created as finite decreasing sequences of numbers under a particular ceiling n. When the ceiling n is finite, there exist as many of these sequences as there are non-empty subsets. In the infinite case, we have to prove that the order type of these finite decreasing sequences is 2^n. We consider the lexicographical ordering.

It is enough to show the following theorem to prove the correctness of the algorithm:

Theorem 3.2 *Let a number $(\rho - 1.m)$ have position α in the system of numbers with the numerus $\rho - 1$. Then there are numbers with numerus $\rho - 1$ that are smaller than $(\rho - 1.m)$ and their order type is α. We can make ordinal numbers with numerus ρ from their mantissae, and these numbers have the order type 2^α if α is infinite. Otherwise, their order type is $2^\alpha - 1$.*

Proof We prove the claim by transfinite induction on α.

Assume that $\alpha = 0$. Hence, the examined number is of the form $(\rho - 1).1$ and there are no smaller numbers with the numerus $\rho - 1$. That is why we are not able to build a number with the numerus ρ of numbers standing before $(\rho - 1).1$. In fact, $2^0 - 1 = 0$.

We are going to use a helpful relation in the following reasoning (preliminary remark). Let us explain it now: Imagine that we have a number $(\rho - 1.\bar{m})$ whose position in the system of numbers with the numerus $\rho - 1$ is γ. It is, of course, possible to make numbers with the numerus ρ from the numbers with the numerus $\rho - 1$ that are smaller than $(\rho - 1.\bar{m})$. Furthermore, we have that the order type of the new numbers is 2^γ or $(2^\gamma - 1)$, depending on whether γ is finite or infinite. Let us denote this order type by δ. Now, let us incorporate the mantissa $\bar{m}$ into the numbers we created in such a way that they remain correct ordinal numbers with the numerus ρ. We know that a new mantissa is made from smaller ones which have to be organized from the greatest to the smallest. It follows that we are obliged to add $\bar{m}$ to the beginning of the existing mantissae:

$$\underbrace{(\rho.m_1), \quad (\rho.m_2) \quad \ldots}_{\text{order type } \delta} \quad \leadsto \quad (\rho.\bar{m}), \quad \underbrace{(\rho.\bar{m}\underbrace{0\ldots0}_{\rho} m_1), \quad (\rho.\bar{m}\underbrace{0\ldots0}_{\rho} m_2)\ldots}_{\text{order type } \delta}$$

Let us continue with the analysis of what α may be. Assume that α is a finite successor ordinal. Then, a number α' exists such that $\alpha = \alpha' + 1$. The number $(\rho - 1.m)$ has a predecessor $(\rho - 1.m')$ whose position in the system of numbers with the numerus $\rho - 1$ is α'. We use the induction hypothesis and the preliminary remark. It is easy to obtain that the order type of the numbers with the numerus ρ that have been built of mantissae smaller than m is

$$(2^{\alpha'} - 1) + 1 + (2^{\alpha'} - 1) = 2^{\alpha'} \cdot 2 - 1 = 2^{\alpha'+1} - 1 = 2^\alpha - 1.$$

Assume that α is an infinite successor ordinal. Then, $\alpha = \alpha' + 1$ and α' is infinite, too. The number $(\rho - 1.m)$ has a predecessor $(\rho - 1.m')$ whose position in the system of numbers with the numerus $\rho - 1$ is α'. The induction hypothesis and the preliminary remark tell us that the order type of the numbers with the numerus ρ that have been built of mantissae smaller than m is

$$2^{\alpha'} + (1 + 2^{\alpha'}) = 2^{\alpha'} + 2^{\alpha'} = 2^{\alpha'} \cdot 2 = 2^{\alpha'+1} = 2^\alpha.$$

Assume that α is a limit ordinal. Then, there exist smaller numbers with the numerus $\rho - 1$ before $(\rho - 1.m)$ whose order type is α:

$$\underbrace{(\rho - 1.m_1), \quad (\rho - 1.m_2), \quad (\rho - 1.m_3), \quad \ldots \quad (\rho - 1.m)}_{\text{order type } \alpha}$$

3.4 An Algorithm for Translating Gentzen's Notation of Ordinal Numbers

Our plan is to make mantissae for numbers with the numerus ρ from them and find out what kind of order type these numbers have, considered altogether as one lexicographically ordered sequence. We know that the induction hypothesis holds for every $\gamma < \alpha$. Based on the preliminary remark, the order type of the numbers with the numerus ρ that have been made from the numbers with the numerus $\rho - 1$ smaller than $(\rho - 1.m)$ is

$$\omega + \sum_{\omega \leq i < \alpha} (1 + 2^i) = \omega + \sum_{\omega \leq i < \alpha} 2^i = \omega + 2^\omega + 2^{\omega+1} + \ldots + 2^{\omega+n} + \ldots$$

The last thing is to calculate the result. We use the lemma:

Lemma 3.2 *Let $\bar{\alpha}$ be an ordinal number in Cantor normal form. Then we have:* $\omega + \sum_{\omega \leq i < \bar{\alpha}} 2^i = 2^{\bar{\alpha}}$.

Proof We show this by transfinite induction on $\bar{\alpha}$.

Let us take the smallest possible $\bar{\alpha} = \omega + 1$. We obtain $\omega + 2^\omega = \omega \cdot 2 = 2^{\omega+1}$. Assume now that $\bar{\alpha}$ is a successor ordinal. It is clear that there is a number δ such that $\delta + 1 = \bar{\alpha}$. We have:

$$\underbrace{\omega + \ldots}_{\text{IH: } 2^\delta} + 2^\delta = 2^\delta \cdot 2 = 2^{\delta+1} = 2^{\bar{\alpha}}$$

Assume that $\bar{\alpha}$ is a limit ordinal. The induction hypotheses tell us that $\omega + \sum_{\omega \leq i < \bar{\alpha}} 2^i = \sup\{2^i, i < \bar{\alpha}\}$. According to the definition of exponentiation, it is equal to $2^{\bar{\alpha}}$.

Now, it is easy to see that $\omega + \sum_{\omega \leq i < \alpha} 2^i = 2^\alpha$ which is our desired result. □

Let us present some examples of how the algorithm works:

Example 3.2 Let us examine the number 2.201. It is built of 1.201 whose position in the system with the numerus 1 is $\omega + 1$. The Cantor normal form that corresponds to 2.201 is:

$$(\omega + 1) + (\omega + \omega) + 2^{\omega+1}$$
$$= \omega \cdot 3 + (2^\omega \cdot 2)$$
$$= \omega \cdot 3 + \omega \cdot 2 = \underline{\omega \cdot 5}$$

Example 3.3 Let us examine the number 2.2001. It is built of 1.2 and 1.1 whose positions in the system with the numerus 1 are ω and 0, respectively. The Cantor normal form that corresponds to 2.2001 is:

$$(\omega + 1) + (\omega + \omega) + (2^\omega + 1) + (2^0 - 1)$$
$$= \omega \cdot 3 + (\omega + 1) = \underline{\omega \cdot 4 + 1}$$

Example 3.4 Let us examine the number 3.2010002001. It is built of 2.201 and 2.2001 whose positions a_1 and a_2 in the system with the numerus 2 have to be determined. We know their corresponding Cantor normal forms, so, we are looking for a_1 and a_2 such that

$$\omega \cdot 3 + a_1 = \omega \cdot 5$$
$$\omega \cdot 3 + a_2 = \omega \cdot 4 + 1$$

It follows that the positions of 2.201 and 2.2001 in the system with the numerus 2 are $a_1 = \omega \cdot 2$ and $a_2 = \omega + 1$, respectively. The Cantor normal form that corresponds to 3.2010002001 is:

$$(\omega + 1) + (\omega + \omega) + (\omega \cdot \omega) + (2^{\omega \cdot 2} + 1) + 2^{\omega+1}$$
$$= \omega^2 + ((2^\omega)^2 + 1) + (2^\omega \cdot 2)$$
$$= \omega^2 + \omega^2 + \omega \cdot 2 = \underline{\omega^2 \cdot 2 + \omega \cdot 2}$$

References

Gentzen, G. 1936. Die Widerspruchsfreiheit der reinen Zahlentheorie. *Mathematische Annalen* 112: 493–565.

Jech, T. 2003. *Set theory*, The Third Millennium Edition. Berlin: Springer.

Chapter 4
Consistency Proof

Abstract Gentzen's consistency proof of 1936 is explained in full detail. The chapter begins with the method of assigning ordinal numbers in Gentzen's representation to derivations. The second part introduces reduction steps for derivations whose endsequent is not in endform and shows, at the same time, that every reduction lowers the ordinal number of the given derivation. The proof consists in analysing cases according to the last inference rule that is used in the derivation to reduce. We obtain the consistency of arithmetic in this way as the derivation of $\to 0 = 1$ would be reduced infinitely many times. Hence, we would construct an infinite decreasing sequence of ordinal numbers which is not possible.

Keywords Consistency of arithmetic · Consistency proof of 1936 · Gentzen's consistency proofs · Reduction steps · Reduction steps for derivations

4.1 How to Assign Ordinal Numbers to Derivations

Definition 4.1 The maximal number of the digits 0 in a row in a mantissa m is the *rank* of m. We often denote the rank of a mantissa by ν.

Example 4.1 The rank of the mantissa 20111101002001110100011 is 3.

The mantissae of ordinal numbers assigned to the derivations have a standard form which we shall denote by *general condition*: The rank of the mantissa must be $\nu > 1$. All parts that are separated from each other by sequences of ν zeros must begin with the digit 2 except for the last part. The last part consists of the digits 1:

$$\rho \, . \, \underbrace{2..0..0}_{\nu} \underbrace{2..0..0}_{\nu} .. \underbrace{0..0}_{\nu} \underbrace{1..1}_{\text{the last part}}$$

If $\rho.m$ is an ordinal number, then it holds that the numerus ρ is greater than or equal to the rank ν of the mantissa m: $\rho \geq \nu$. See Example 3.1.

Definition 4.2 Let $\rho.m$ be an ordinal number. Let the rank of the mantissa m be ν. The difference between ρ and ν is called $excess_{\rho.m}$ (Überschuß).

Example 4.2 Let us take an ordinal number $\alpha = 5.201110100200110100011$. The $excess_\alpha$ is 2.

We often work with derivations of the premises of some rule of inference. These derivations will, of course, have ordinal numbers, too. As we often talk about these ordinal numbers, we need to introduce a short name for them.

Definition 4.3 The ordinal number of a derivation whose endsequent is a premise of an inference rule is called *parental number* of this rule. The mantissa and the numerus of this ordinal number are called *parental mantissa* and *parental numerus*, respectively.

We shall see that there is at least one sequence of 0's in the mantissa of every assigned ordinal number. It means that the mantissa does not consist only of its final part including exclusively 1's. Further, we see that no derivation gets assigned an ordinal number with the numerus 0 or 1, because the rank of the mantissa of every assigned number must be at least 2.

We proceed to explain how to assign the ordinal numbers to derivations.

Assume the endsequent of the derivation is an *initial sequent*. Such a derivation gets assigned the ordinal number $2.2001..1$ where the exact number of the digits 1 in the last part is one greater than the number of the logical operations in the initial sequent. This ordinal number satisfies the general condition and it is a correct ordinal number according to Definition 3.1: It is built of 1.2 and $1.1..1$. These two are built of 0.2 and $0.1..1$.

Assume the endsequent is derived by the *negation rule* $\frac{\Gamma, A \to 0=1}{\Gamma \to \neg A}$ or by $\forall$-*introduction* $\frac{\Gamma \to F(a)}{\Gamma \to \forall x F(x)}$ where a is an eigenvariable. Assume α is the parental number. The ordinal number of the whole derivation is built of α by adding the digit 1 to the end of α.

Example 4.3

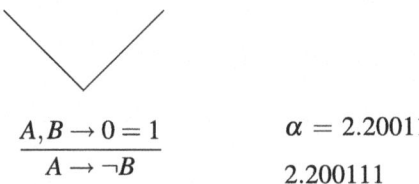

$$A, B \to 0 = 1 \qquad \alpha = 2.20011$$
$$\overline{A \to \neg B} \qquad\qquad 2.200111$$

Adding the digit 1 can be interpreted intuitively as the representation of the logical operation that was added to the endsequent after the use of the rule. The ordinal number of the whole derivation satisfies the general condition because α itself has satisfied it. It is also a correct ordinal number: The number α is correct by assumption.

4.1 How to Assign Ordinal Numbers to Derivations

Every sequence of 0's in a mantissa divides this mantissa into the mantissae of ordinal numbers with a smaller numerus. So, the final part that consists of the digits 1 must be the mantissa of an ordinal number, too. We do not need to examine what kind of ordinal numbers allow of such a mantissa because the mantissa that consists of 1's builds a correct ordinal number with an arbitrary numerus.

Assume the endsequent is derived by the *chain rule* and the derivation of every premise has its own ordinal number: $\rho_1.m_1 .. \rho_k.m_k$. We are going to work with all parental mantissae: $m_1, m_2 .. m_k$. First, we choose such a parental mantissa that has the maximal rank ν. If there are some identical parental mantissae, that belong to the derivations of different premises, we have to distinguish these mantissae by adding the following sections to their ends:

- $\nu + 1$ zeros and 1
- $\nu + 1$ zeros and 11
- $\nu + 1$ zeros and 111
- ..

Every mantissa that needs to be distinguished gets exactly one section so that these mantissae will differ in the number of 1's at their ends. All parental mantissae are *different* now. The modified mantissae may be improper mantissae for their numeri because the modified mantissa and the original numerus do not have to build a correct ordinal number together. It does not matter at this point. The modified mantissae are correct mantissae as there exists a numerus for each of them such that they would build a correct ordinal number together. The rank of every modified mantissa is $\nu+1$. The section of 0's that is $\nu + 1$ in length occurs there exactly once, so, they are built of two smaller sub-mantissae. It is certain that the sub-mantissa that forms the initial part is greater than the sub-mantissa 1 .. 1 that forms the final part because the initial part starts with the digit 2 according to the general condition applied to the parental mantissae.

Second, we order all parental mantissae (adjusted when necessary) according to their values: from the greatest to the smallest. Now, we connect them by sequences of 0's that are $\nu + 2$ in length and add an additional sequence of 0's followed by the digit 1 to the end:

$$\underbrace{m_{i_1} > m_{i_2} > m_{i_3} .. > m_{i_k}}_{\text{the adjusted parental mantissae ordered form the greatest to the smallest}}$$

The mantissa m for the whole derivation:

$$m_{i_1} \underbrace{0..0}_{\nu+2} m_{i_2} \underbrace{0..0}_{\nu+2} .. m_{i_k} \underbrace{0..0}_{\nu+2} 1$$

We go on to define a numerus ρ for the whole derivation. To define this, we use the notion of excess (Definition 4.2). The numerus ρ is *the smallest natural number* that satisfies these three conditions:

- $\rho \geq \nu + 2$ where $\nu + 2$ is the rank of the mantissa m defined earlier.

44   4 Consistency Proof

- Let us take excess$_{\rho_1.m_1}$ ..excess$_{\rho_k.m_k}$ where $\rho_1.m_1$ ..$\rho_k.m_k$ are all parental numbers. Assume the maximal excess among these values is excess$_{\rho_i.m_i}$ where $i \leq k$. The numerus ρ, which we are looking for, must satisfy this condition: $\rho - (\nu+2) \geq$ excess$_{\rho_i.m_i} - 2$. It means that excess$_{\rho.m}$ must be at least (excess$_{\rho_i.m_i} - 2$) where $\rho.m$ stands for the future ordinal number of the whole derivation. We can also write it this way: $\rho \geq \nu +$ excess$_{\rho_i.m_i}$.
- Let us take all succedent formulas of the premises that stand on the left of the main premise. We choose the one that contains a maximal number of logical operations (every occurrence counts). Assume the maximal number found is y. Then, the numerus ρ must satisfy this condition: $\rho - (\nu + 2) \geq 2y$. It means that excess$_{\rho.m}$ must be at least $2y$ where $\rho.m$ stands for the future ordinal number of the whole derivation. We can also write it this way: $\rho \geq \nu + 2(1 + y)$.

Finally, we connect the numerus ρ, which we found, with the defined mantissa m and obtain the ordinal number of the whole derivation. The mantissa m satisfies the general condition because $\nu + 2 > 1$ and every part separated by the sequence of 0's that is $\nu + 2$ in length begins with the digit 2. It is clear because these parts were independent mantissae. The last part contains only 1's.

The number we created is a correct ordinal number: The numerus ρ is greater than or equal to the rank $\nu + 2$ of the mantissa m. The sub-mantissae used for building the whole mantissa m are correct mantissae of smaller ordinal numbers. It means that ordinal numbers with the numerus $\nu + 1$ can be built of them. We also ensured that the sub-mantissae are all different and ordered according to their values.

Assume the endsequent is derived by the *induction rule* $\dfrac{\Gamma \to F(0) \quad F(a), \Delta \to F(a+1)}{\Gamma, \Delta \to F(t)}$ where a is an eigenvariable. The mantissa m for the whole derivation looks like this:

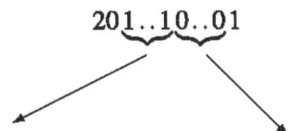

We put here exactly as many digits 1 as the sequence of 1's at the corresponding position in the greater parental mantissa has, plus one more.	Assume that ν is the *maximal* rank of the parental mantissae. We put here $\nu + 2$ occurrences of the digit 0.

A comment on the instruction how to determine the length of the sequence of 1's: Assume that the *greater* parental mantissa is m_1. If m_1 begins with the sequence 200, we insert there exactly one digit 1. If m_1 begins with the sequence 201..1, we insert there one digit 1 more than the mentioned sequence has. For example, if m_1 were 20111101002001110100011, we would insert there exactly five digits 1. The mantissa m_1 cannot begin with the sequence 202 because each mantissa must be built of different sub-mantissae. It cannot begin with the sequence $20x$ where $x \geq 3$ because the digits 3 and greater are not allowed. It cannot begin with 21 either because the digits 2 and 1 never stand immediately by each other in a mantissa. See Example 3.1.

4.1 How to Assign Ordinal Numbers to Derivations   45

We know that $\nu + 2 > 1$, and it is the rank of the new-built mantissa m. The first section begins with the digit 2 and the final section consists of the digit 1. That is why the mantissa m satisfies the general condition.

The numerus ρ is *the smallest natural number* that satisfies these three conditions:

- $\rho \geq \nu + 2$
- The excess of the ordinal number of the whole derivation must be at least (excess$_{\rho_i.m_i} - 2$) where $\rho_i.m_i$ is the parental number with the maximal excess. So, once more, we need a numerus ρ such that $\rho - (\nu + 2) \geq$ (excess$_{\rho_i.m_i} - 2$).
- Assume the number of the logical operations in the formula $F(0)$ is y. Then, we have to choose a numerus ρ such that $\rho - (\nu + 2) \geq 2y$.

We have defined a mantissa and a numerus for the whole derivation. They build a correct ordinal number together: The value of the numerus ρ is at least as large as the value of the rank $\nu + 2$ of the mantissa m. The mantissa m is built of two sub-mantissae: 201..1 and 1. The first one is a mantissa for the numerus 1 (the smallest allowed numerus in this case) and the second one is a mantissa for the numerus 0 (the smallest numerus ever). Good news for us because the numerus ρ must be at least $\nu + 2$ where $\nu > 1$. We know that the numbers with the numerus ρ have to be built of numbers with the numerus $\rho - 1$. According to our analysis, both mentioned sub-mantissae are proper mantissae for such numbers.

4.2 Lowering the Ordinal Numbers After Reduction Steps for Derivations

Let us summarize what we have done until now: We took an arbitrary old derivation in PA (Sects. 2.1–2.3) and transformed it into a new derivation (Sects. 2.7, 2.8). Then, we replaced all free variables except eigenvariables by arbitrary numerals. After the values of the terms had been calculated, we replaced these terms by the numerals which have the corresponding values. The derivation modified this way will be used and reduced further below. Furthermore, we have explained how to assign an ordinal number to a derivation.

Now, we are going to define reduction steps for derivations and, at the same time, we show that these reduction steps lower the ordinal numbers of the derivations. This is our theorem to prove:

Theorem 4.1 *Each reduction step for derivations changes the derivation to another correct derivation and the endsequent remains the same or is reduced according to exactly one reduction step for sequents. The numerus does not get greater and the mantissa gets smaller after the reduction step. It means that the whole ordinal number goes down. The rank of the mantissa does not change except for the case e3. In this case, it increases exactly by 2.*

If the endsequent is in endform, no reduction step is defined for its derivation. Therefore, we are allowed to assume that the derivation, which we work with, has an endsequent that is not in endform. The proof of Theorem 4.1 will be completed at the end of this section. Let us begin:

(**a**) Assume the endsequent of the derivation is an *initial sequent*. We carry out one step according to the instructions for reducing the initial sequents (Sect. 2.6). We obtain an initial sequent, hence, we have a correct derivation again. One logical operation disappears after such a reduction step. Even steps 13.5 are used in the way that the affected formula is erased and never stays among the antecedent formulas. The ordinal number of the examined derivation is calculated according to the number of the logical operations in the initial sequent:

$$\underbrace{A\&B \to A\&B}\qquad \underset{\leadsto}{13.2} \qquad \underbrace{A\&B \to A}$$

Assume there are no logical operations in the formulas A and B. Therefore, the ordinal number of this derivation is 2.200111.

The ordinal number of this derivation is 2.20011. The numerus and the rank of the mantissa are unchanged. The mantissa is smaller in comparison with the case before the reduction. So, the whole number is smaller.

(**b**) Assume the endsequent is derived by ∀-*introduction*:

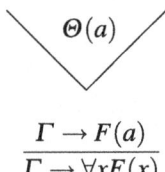

$$\frac{\Gamma \to F(a)}{\Gamma \to \forall x F(x)}$$

Here a is an eigenvariable and does not occur in Γ, $\forall x F(x)$. The reduction consists in omitting the endsequent and in replacing the eigenvariable a with an arbitrary numeral $\bar{n}$ in the whole derivation $\Theta(a)$. The premise $\Gamma \to F(\bar{n})$ becomes a new endsequent.

We obtain:

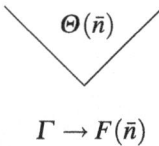

$$\Gamma \to F(\bar{n})$$

Now, we have to calculate the values of all terms that do not contain any variables after the replacement and write the numerals with the corresponding values instead. The reduced derivation is correct and the endsequent was reduced according to step 13.21.

Let us look at the ordinal numbers matter. Assume that the ordinal number of the derivation of the sequent $\Gamma \to F(a)$ is $\rho.m$. It means that the ordinal number of the whole original derivation is $\rho.m1$. We simply added the digit 1 to the end of

4.2 Lowering the Ordinal Numbers After Reduction Steps for Derivations

the mantissa m. The derivation of $\Gamma \to F(\bar{n})$ has the same ordinal number as the original derivation of the premise $\Gamma \to F(a)$, hence $\rho.m$. We see that the numerus remained unchanged, as well as the rank, but the mantissa became smaller after the reduction.

(c) Assume the endsequent is derived by the *negation rule*:

$$\frac{\Gamma, A \to 0 = 1}{\Gamma \to \neg A}\ \Theta$$

The reduction transforms the derivation in the same way as in item (b):

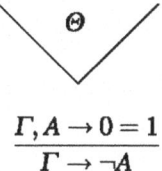

$$\frac{\Gamma, A \to 0 = 1}{\Gamma \to \neg A}\ \Theta \quad \overset{13.23}{\leadsto} \quad \Gamma, A \to 0 = 1\ \Theta$$

We leave out the endsequent and take its premise as a new one. It is obvious that if we reduce the original endsequent according to 13.23, we obtain exactly the sequent we have as a new endsequent. The analysis of the assigned ordinal numbers is the same as in item (b).

(d) Assume the endsequent is derived by the *induction rule*:

$$\frac{\Gamma \to F(0) \quad F(a), \Delta \to F(a+1)}{\Gamma, \Delta \to F(t)}\ \Psi\ \Theta(a)$$

Here a is an eigenvariable and does not occur in Γ, Δ, $F(0)$, $F(t)$. We know that the term t is a numeral because we replaced all free variables except eigenvariables by arbitrary numerals. Furthermore, we calculated the values of such terms that did not contain any free variables and replaced these terms with the corresponding numerals. The term t occurs in the endsequent, so, it is impossible that it contains any eigenvariable. Assume that t is $\bar{n} > 0$ and $\bar{m} = \bar{n} - 1$. We create derivations $\Theta(\bar{0}), \Theta(\bar{1}) \ldots \Theta(\bar{m})$ that are similar to the derivation $\Theta(a)$. The difference is that the variable a is replaced with the numeral in parentheses; and again, the values of the terms are calculated if possible. The whole reduced derivation has this form:

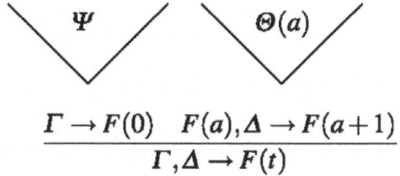

48   4 Consistency Proof

The numbers situated next to the sequents represent the ordinal numbers assigned to the derivations of these sequents. The induction rule has become the chain rule and the endsequents are the same in both cases: $\Gamma, \Delta \to F(\bar{n})$.

Let us realize what the new numerus $\bar{\rho}$ looks like:

- The numerus $\bar{\rho}$ depends on the maximal rank of the parental mantissae. Assume it is v. The maximal rank of the parental mantissae in the original derivation is the same because both derivations have the same set of parental mantissae: $\{m_1, m_2\}$. Although the last rule in the original derivation is the induction rule, the numerus ρ depends on v too and the corresponding condition for both numeri is the same: $\rho \geq v + 2, \bar{\rho} \geq v + 2$.
- The numerus $\bar{\rho}$ depends on the maximal excess of the parental numbers. Let us denote it by $\text{excess}_{\rho_i.m_i}$. The set of parental numbers in both derivations is the same: $\{\rho_1.m_1, \rho_2.m_2\}$. It follows that the condition for ρ and $\bar{\rho}$ concerning the maximal excess of the parental numbers is the same: $\rho \geq \text{excess}_{\rho_i.m_i} + v$, $\bar{\rho} \geq \text{excess}_{\rho_i.m_i} + v$.
- The numerus $\bar{\rho}$ depends on the maximal number of the logical operations in a succedent formula of some premise that stands on the left of the main premise. All succedent formulas contain $|F(0)|$ logical operations. The original numerus ρ depends on exactly the same value.

A numerus is the smallest natural number that satisfies three conditions (Sect. 4.1) given by three values analysed above. We found out that these three significant values are the same in the original as well as in the reduced derivation. This results in $\rho = \bar{\rho}$.

Let us calculate the new mantissa $\bar{m}$. The old mantissa m is of the form:

$$m = 20\underbrace{1..1}\underbrace{0..0}1$$
$$v+2$$

Assume m_1 is greater than m_2. We put here as many digits 1 as m_1 has at the corresponding position, plus one more.

The new mantissa $\bar{m}$ is built of the parental mantissae. We see that the parental mantissa m_2 occurs there n times, so, we are obliged to distinguish its different occurrences:

$$m'_2 = m_2 \underbrace{0..0}_{v+1} 1$$
$$m''_2 = m_2 \underbrace{0..0}_{v+1} 11$$
$$\vdots$$
$$m_2 \underbrace{',..,'}_{n \text{ commas}} = m_2 \underbrace{0..0}_{v+1} \underbrace{1..1}_{n}$$

We assumed that $m_1 > m_2$. This holds after the transformation of the mantissa m_2, too: $m_1 > m'_2, m_1 > m''_2 .. m_1 > m_2^{\cdot\cdot'}$. Let us explain this. When we consider the

4.2 Lowering the Ordinal Numbers After Reduction Steps for Derivations

assumption $m_1 > m_2$, then two possibilities can occur: (1) Either the first digit that m_1 and m_2 differ in is smaller in m_2 or (2) m_2 is an initial part of m_1. In the first case, it is easy to see that m_1 has the desired relation with every $m_2'^{..'}$. Let us look at the second case. The initial part of every mantissa $m_2'^{..'}$ is the mantissa m_2 followed by a section of $\nu + 1$ zeros. As we assumed that the maximal rank of both m_1 and m_2 is ν, the mantissa m_1 can have at most ν zeros after its initial part m_2. We know that m_1 does not end with the digit 0. In the worst case, m_1 and an arbitrary $m_2'^{..'}$ start to differ in the digit at the position $\nu + 1$ of the examined section of 0's. They differ in such a way that the digit at the particular position in m_1 is greater than the digit at the same position in $m_2'^{..'}$ because there is no section of $\nu + 1$ zeros in m_1.

The new mantissa $\bar{m}$ looks like this:

$$\bar{m} = m_1 \underbrace{0..0}_{\nu+2} m_2'^{..'} \underbrace{0..0}_{\nu+2} ..m_2'' \underbrace{0..0}_{\nu+2} m_2' \underbrace{0..0}_{\nu+2} 1$$

We need to show that the first digit that m and $\bar{m}$ differ in is smaller in $\bar{m}$. The new mantissa $\bar{m}$ begins with the larger parental mantissa m_1 that has the form 201..1.. in general. If we compare it with m, we find out that $\bar{m}$ has one digit 1 less in its first section of 1's. The result is that although $\bar{m}$ may be much longer than m, $\bar{m}$ is smaller than m.

Let us repeat the relation between the ordinal numbers of the original and the reduced derivation. We mean, between the numbers $\rho.m$ and $\bar{\rho}.\bar{m}$:

- $\rho = \bar{\rho}$
- $m > \bar{m}$
- The rank of m and $\bar{m}$ is the same: $\nu + 2$.

Let us examine the case when $\bar{n} = 0$:

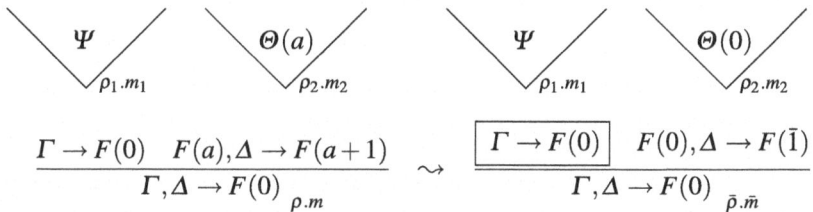

The induction rule turns into the chain rule again. The second premise $F(0), \Delta \to F(\bar{1})$ of the chain rule is not used. It stands behind the main premise. The list of formulas Δ is among the antecedent formulas of the new endsequent because we decided to put it there. The chain rule allows us to do so and we need it, of course, to obtain the same endsequent as in the original derivation.

Let us compare the numeri ρ and $\bar{\rho}$:

- The maximal rank of the parental mantissae is the same in both derivations. Let us denote it by ν.
- The maximal excess of the parental numbers is the same in both derivations.

- Assume the formula $F(0)$ contains y logical operations. We have $\rho \geq \nu + 2 + 2y$. Since there are no premises on the left of the main premise, we have $\bar{\rho} \geq \nu + 2 + 0$. Therefore, it can occur that $\bar{\rho} \leq \rho$. This is in harmony with our statement in Theorem 4.1.

The mantissae m and $\bar{m}$ are both made from m_1 and m_2. The mantissa m is made according to the instructions described in the section for the induction rule, so, it contains one digit 1 more in its first section of 1's than the mantissa $\bar{m}$. The mantissa $\bar{m}$ is made according to the instructions for the chain rule. These instructions tell us to order the *unchanged* parental mantissae m_1, m_2 and connect them in order to build $\bar{m}$. The outcome is $\bar{m} < m$.

(e) Assume the endsequent is derived by the *chain rule*. First, we apply the following *preparation step*:

If the succedent formula of the main premise is a false atomic sentence, we have to choose another sequent from the premises as the main premise: We take such a sequent that is the first one from the left whose succedent formula is a false atomic sentence, too. This does not increase the ordinal number of the derivation. The mantissa does not change for sure, because it was built of all parental mantissae and they are untouched. It is possible that the numerus decreases because we do not take into account some succedent formulas while selecting the one with the maximal number of the logical operations: We might have fixed a new main premise which is on the left of the old main premise, so, fewer formulas are available to choose from. The chosen formula could contain fewer logical operations than the formula chosen before the preparation step was applied.

It can happen that the chain rule uses the option of adding additional formulas to the antecedent of the conclusion after the preparation step to obtain the same conclusion as before the step. This is possible and correct according to Lemma 2.3.

Lemma 4.1 *If the endsequent is not in endform, the main premise of the chain rule that the endsequent was derived by is not in endform either.*

Proof Let us have a derivation whose endsequent is not in endform and whose last rule of inference is the chain rule. If the main premise were in endform, the succedent formula of the main premise would be either a true atomic sentence or it would be a false atomic sentence and there would be at least one false atomic sentence among the antecedent formulas. As far as the first case is concerned, it is obvious that the endsequent would be in endform too because its succedent formula is identical to the one from the main premise. This contradicts the assumption. An explanation for the second case: The main premise is the first premise in a row whose succedent formula is a false atomic sentence. We ensured this by applying the preparation step. Thus, the false atomic sentence situated in the antecedent of the main premise must have got into the antecedent of the endsequent. It could not have been deleted. Hence, the endsequent would be in endform and we would obtain a contradiction again. □

4.2 Lowering the Ordinal Numbers After Reduction Steps for Derivations 51

We showed that if the endsequent is not in endform, the main premise is not in endform either. Thus, our induction hypothesis tells us that there exists a reduction step for the derivation of the main premise.

Now, we are about to analyse *seven* cases. Sentences with which every item begins are the following: 'The endsequent is derived by the chain rule. The main premise is not in endform, so there exists a reduction step for its derivation.' Then it continues:

e1: Assume the main premise is transformed according to reduction step 13.2 while reducing its derivation. It is possible to apply step 13.2 to the endsequent too, because they have the same succedent.

e2: Assume the main premise is transformed according to reduction step 13.5 while reducing its derivation. There is the same formula as the affected one in the antecedent of the endsequent. It is therefore possible to apply step 13.5 to the endsequent, too.

e3: Assume the main premise is transformed according to reduction step 13.5 while reducing its derivation and the affected formula B is not among the antecedent formulas of the endsequent. It was deleted because it is the succedent formula of a premise that stands on the left of the main premise. Since B was used in reduction step 13.5, it is not atomic. So, by the induction hypothesis, a reduction step for the derivation of the premise $\Gamma \to B$ that B comes from is defined. Assume step 13.2 is applied to this premise while reducing its derivation.

e4: Assume the main premise remains unchanged after the reduction step was applied to its derivation and the last rule of inference used in its derivation is a chain rule that needs reduction treated in item e3. (As we shall see, the reduction described in e3 does not change the endsequent.)

e5: Assume the main premise is transformed according to reduction step 13.5 while reducing its derivation and the affected formula B is not among the antecedent formulas of the endsequent. It was deleted because it is the succedent formula of a premise that stands on the left of the main premise. Since B was used in reduction step 13.5, it is not atomic. So, by the induction hypothesis, a reduction step for the derivation of the premise $\Gamma \to B$ that B comes from is defined. (Note that this case is identical to case e3 until now.) Assume the premise $\Gamma \to B$ remains unchanged after the reduction step was applied to its derivation and the last rule of inference used in its derivation is a chain rule that needs reduction treated in item e3.

e6: Assume the main premise remains unchanged after the reduction step was applied to its derivation and the last rule of inference used in its derivation *is not a chain rule that needs reduction treated in item e3*. (Note that this is similar to e4, but there is a difference in the rule of inference applied in order to derive the main premise.)

e7: Assume the main premise is transformed according to reduction step 13.5 while reducing its derivation and the affected formula B is not among the antecedent formulas of the endsequent. It was deleted because it is the succedent formula of a premise that stands on the left of the main premise. Since B was used in reduction step 13.5, it is not atomic. So, by the induction hypothesis, a reduction step for the derivation of the premise $\Gamma \to B$ that B comes from is defined. Assume the premise $\Gamma \to B$ remains unchanged after the reduction step was applied to its derivation and the last rule of inference used in its derivation *is not a chain rule that needs reduction treated in item e3*. (Note that this is similar to e5.)

The following scheme may be more transparent:

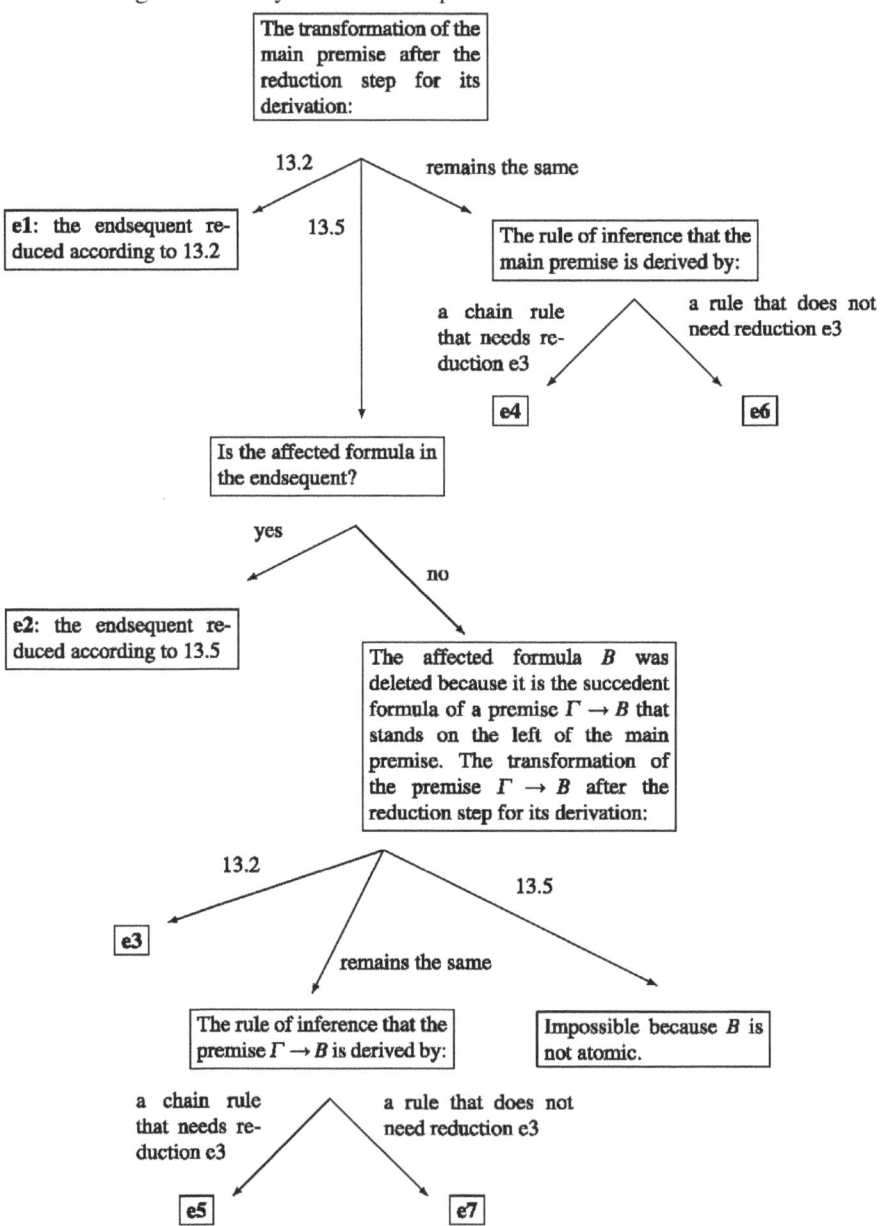

4.2 Lowering the Ordinal Numbers After Reduction Steps for Derivations

The following analyses of each case will start by repeating the corresponding description to make it more comfortable to read.

(**e1**) The endsequent is derived by the chain rule. The main premise is not in endform, so there exists a reduction step for its derivation. Assume the main premise is transformed according to reduction step 13.2 while reducing its derivation. It is possible to apply step 13.2 to the endsequent too, because they have the same succedent:

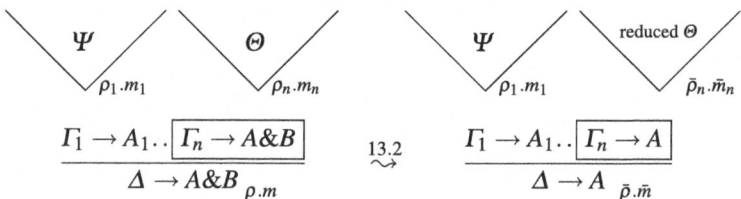

Picture e1

The chain rule is correct and the endsequent, compared to the old one, is reduced according to step 13.2. If necessary, we have to make the same choice as while reducing the derivation of the main premise.

Lemma e1 *Reduction e1 does not change the numerus and the rank of the mantissa. The ordinal number of the reduced derivation is smaller than that of the original derivation. It means that the following applies to the ordinal numbers in Picture e1: $\rho = \bar{\rho}$, $\bar{m} < m$, $\bar{m}$ and m have the same rank.*

Proof We are going to prove this by induction. Let us take an inference step in our derivation that uses a chain rule which needs to be reduced according to e1 and there is *not* such an inference step above the chosen step that has to be reduced according to e1, too. Assume the chosen step is of the form as displayed in Picture e1.

We see that the main premise $\Gamma_n \rightarrow A\&B$ was modified while reducing its derivation, concretely, step 13.2 was applied. It helps us to deduce that the original main premise $\Gamma_n \rightarrow A\&B$ must have been derived by one of these rules:

- It is an initial sequent.
- ∀-introduction
- negation rule
- Chain rule that needs reduction according to e1.

The reductions for the derivations whose last inference rule is different from all rules mentioned above do not change the endsequent or they modify it according to step 13.5. Since we assumed that we have chosen the first use of the chain rule that needs reduction according to e1, we obtain that one of the first three possibilities must be the right one (initial sequent, ∀-introduction, negation rule). So, we look at the reductions of these rules and it is easy to find out that the following holds:

- $\bar{\rho}_n = \rho_n$
- $\bar{m}_n < m_n$

- The mantissae $\bar{m}_n$ and m_n have the same rank.

Now, we are going to analyse the relation between $\rho.m$ and $\bar{\rho}.\bar{m}$, what is actually the main thing we are interested in:

The numeri ρ and $\bar{\rho}$ are equal. Their values depend on these three quantities: the maximal rank of the parental mantissae, the maximal excess of the parental numbers, the maximal amount of logical operations in the succedent of a premise that stands on the left of the main premise. They are the same in the original as well as in the reduced derivation.

Since the mantissae m_n and $\bar{m}_n$ have the same rank, we obtain that m and $\bar{m}$ have the same rank, too. Let us compare their values. Assume we have ordered the parental mantissae of the original derivation:

$$k_1 > k_2 > .. k_n$$

Similarly, we have ordered the parental mantissae of the reduced derivation:

$$k'_1 > k'_2 > .. k'_n$$

We know that there is an index i such that $k_i > k'_i$. The mantissae m and $\bar{m}$ are built of $k_1 .. k_n$ and $k'_1 .. k'_n$, respectively. So, the first digit they differ in comes from this ith section and it is (almost) clear that the specific digit is smaller in $\bar{m}$ than in m. We stop for a while and think a little bit deeper about why this holds in the case when k'_i is an initial part of k_i. Let us call the maximal rank of the parental mantissae ν, no matter which derivation we consider. The way the mantissa $\bar{m}$ is created reveals that k'_i, as a part of $\bar{m}$, is followed by $\nu + 1$ or $\nu + 2$ zeros. When we take account of the fact that there can be sections of zeros in k_i whose maximal length is ν and k_i does not end with 0 for sure, then it is easy to verify $\bar{m} < m$.

There is another interesting situation that can occur. Assume there was a group of the parental mantissae in the original derivation that were all identical to m_n. They must have been distinguished by adding $\nu + 1$ zeros and some 1's to their ends before we have built m of them. The group of these identical mantissae lost one member after $\bar{m}_n$ had been substituted for m_n. The modified group was used for building $\bar{m}$. Hence, the members must have been distinguished, too. This time, each member got one digit 1 less. Note that the old mantissa m and the new mantissa $\bar{m}$ differ in this digit. Whereas the old mantissa m still contains 1 at the end of the first modified sub-mantissa, $\bar{m}$ has 0 at this position.

Induction hypothesis: For the ordinal number of a derivation whose endsequent is derived by the chain rule that needs reduction e1, we have that it gets smaller after the reduction step. The reduction step makes it smaller because the new mantissa is smaller than the old one. Furthermore, the numerus and the rank remain the same.

We need to show that this holds for the next sequent derived by the chain rule that also needs reduction e1. The analysis would look like the analysis above that led to the induction hypothesis. □

4.2 Lowering the Ordinal Numbers After Reduction Steps for Derivations

(**e2**) The endsequent is derived by the chain rule. The main premise is not in endform, so there exists a reduction step for its derivation. Assume the main premise is transformed according to reduction step 13.5 while reducing its derivation. There is the same formula as the affected one in the antecedent of the endsequent. It is therefore possible to apply step 13.5 to the endsequent, too. We will do it and use, of course, the same formula as the one affected by the reduction of the derivation of the main premise. If necessary, we have to make the same choice as while reducing the derivation of the main premise.

Now, it is our turn to make the reduced derivation correct again. Let us examine some situations that may occur.

1. The analysed reduction looks like this in general:

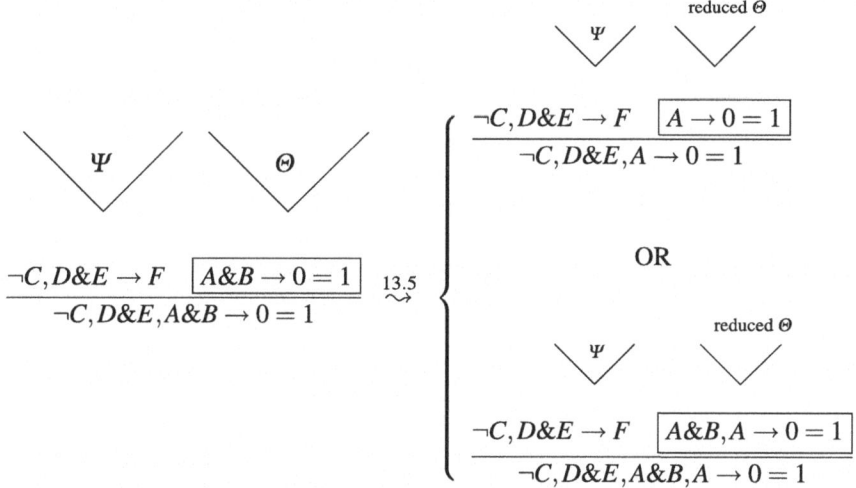

There are two possible ways of reducing the derivation of the main premise. If the affected formula $A\&B$ is deleted after the reduction step, we preserve this while reducing the endsequent. Similarly, we will not erase it if the affected formula remains among the antecedent formulas of the main premise. The chain rule is correct in both cases and the endsequent is obviously reduced according to step 13.5.

2. The formula, in this example A, that came into existence because of the reduction is the succedent of a premise that stands on the left of the main premise:

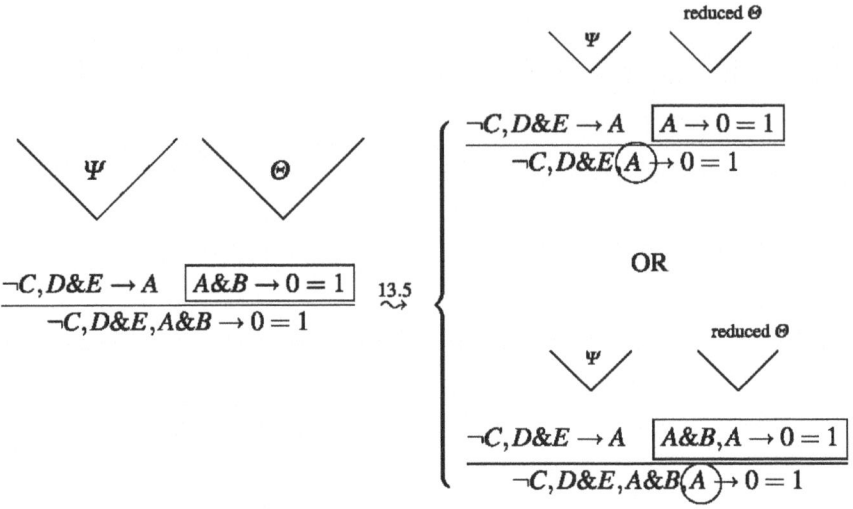

On the one hand, the chain rule orders us not to record the formula A (in the circle). On the other hand, it allows us to add an arbitrary formula to the antecedent. Hence, the chain rule remains correct and the endsequent is reduced according to 13.5.

3. The formula, in this example A, that came into existence because of the reduction is among the antecedent formulas of a premise that stands on the left of the main premise:

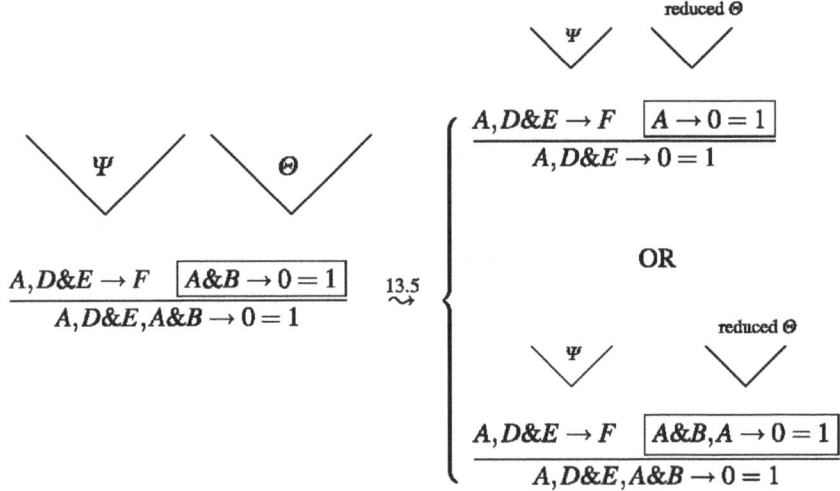

Both the chain rule and reduction step 13.5 tell us not to record the formula A among the antecedent formulas of the endsequent because the same formula is already there. So, there is just one copy of A in the antecedent of the endsequent. The chain rule remains correct and the endsequent is reduced according to 13.5.

4.2 Lowering the Ordinal Numbers After Reduction Steps for Derivations

4. The affected formula $A\&B$ that occurs in the antecedent of the main premise does not get into the antecedent of the endsequent because another copy of the same formula is in the antecedent of a premise that stands on the left of the main premise:

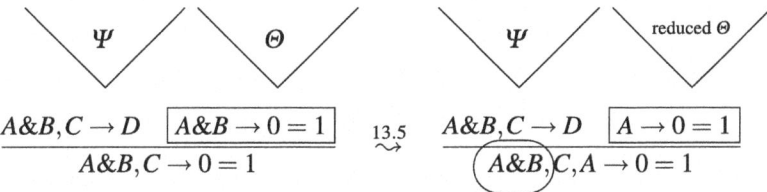

Now this is the point when we take advantage of being allowed to let the affected formula stay among the antecedent formulas. The circled formula is the one which was used while reducing the endsequent. Note that although the formula was erased while reducing the main premise, we let it be after the reduction was applied to the endsequent. The chain rule remains correct and the endsequent is reduced according to 13.5.

To proceed further, we present a lemma that is very similar to Lemma e1, but holds for this particular case:

Lemma e2 *Reduction e2 does not change the numerus and the rank of the mantissa. The ordinal number of the reduced derivation is smaller than that of the original derivation because the new mantissa is smaller than the old one.*

Proof We are going to prove this by induction. Let us take an inference step in our derivation that uses a chain rule which needs reduction according to e2 and there is *not* such an inference step above the chosen step that has to be reduced according to e2, too. Assume the chosen step is of the form:

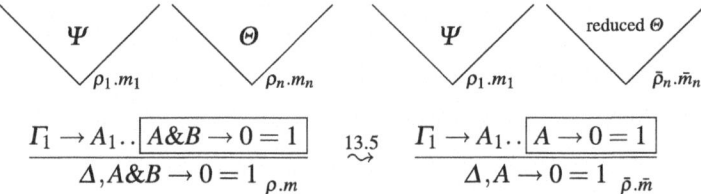

The fact whether the affected formula remains in the antecedent of the endsequent has no influence on the ordinal number of the reduced derivation. The calculation of the new ordinal number $\bar{\rho}.\bar{m}$ is based on the parental numbers and these are set.

We see that the main premise $A\&B \to 0 = 1$ was modified while reducing its derivation, concretely, step 13.5 was applied. It helps us to deduce that the original main premise $A\&B \to 0 = 1$ must have been derived by one of these rules:

- It is an initial sequent.
- Chain rule that needs reduction according to e2.

The reductions for the derivations whose last inference rule is different from all rules mentioned above do not change the endsequent or they modify it according to step 13.2. The sequent $A \& B \to 0 = 1$ must be an initial sequent. The second mentioned possibility would contradict our choice of the examined inference step we made at the beginning of the proof.

It follows:

- $\bar{\rho}_n = \rho_n$
- $\bar{m}_n < m_n$
- The mantissae $\bar{m}_n$ and m_n have the same rank.

We want to compare $\rho.m$ and $\bar{\rho}.\bar{m}$. The numeri ρ and $\bar{\rho}$ are equal. The rank of the mantissae m and $\bar{m}$ is the same and we have $\bar{m} < m$. An explanation for this would be the same as the one we provided in Lemma e1. Actually, the whole proof continues the same way as the one of Lemma e1. □

(e3) Hauptfall[1]: The endsequent is derived by the chain rule. The main premise is not in endform, so there exists a reduction step for its derivation. Assume the main premise is transformed according to reduction step 13.5 while reducing its derivation and the affected formula B is not among the antecedent formulas of the endsequent. It was deleted because it is the succedent formula of a premise that stands on the left of the main premise:

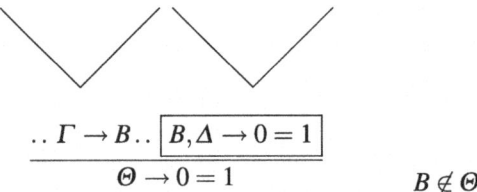

Since B was used in reduction step 13.5, it is not atomic. So, by the induction hypothesis, a reduction step for the derivation of the premise $\Gamma \to B$ is defined. Assume step 13.2 is applied to this premise while reducing its derivation.

There are three similar cases to distinguish:

1. $B = \forall x F(x)$:

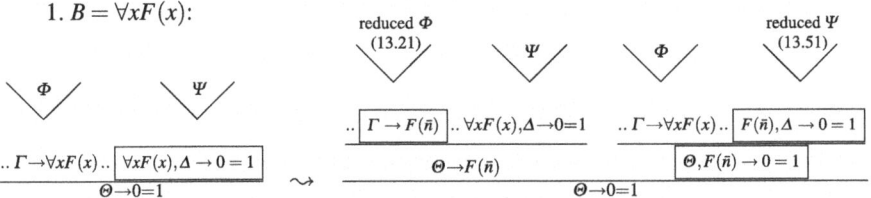

We have the derivation Ψ of the main premise and we know that we are able to reduce it. This reduction has transformed the main premise according to step 13.51. There is a numeral $\bar{n}$ in the endsequent of the reduced derivation Ψ instead of the bound variable x. The same numeral has to be chosen to transform the sequent $\Gamma \to \forall x F(x)$ according to step 13.21 while reducing the derivation Φ. The endsequents of the reduced derivations Φ and Ψ are used as main premises in two independent

---

[1] Gentzen's name for this case.

4.2 Lowering the Ordinal Numbers After Reduction Steps for Derivations

supporting chain rules, let us call them the left one and the right one. Note that these chain rules resemble the last inference step in the original derivation. The conclusions of the supporting chain rules are connected with the help of the third chain rule and we obtain the same endsequent as in the original derivation.

The left chain rule could have been carried out because of Lemma 2.3 which claims that if we take a sequent standing on the left of the main premise and make it a new main premise, the antecedent of the new conclusion will be a subset of the antecedent of the old conclusion. To make sure that all formulas from Θ are there in the antecedent of the conclusion of the left rule, we simply add missing formulas there. The definition of the chain rule allows us to do so.

A commentary on the right rule follows. If the formula $\forall x F(x)$ affected by reduction step 13.51 had not disappeared after this step, it would not concern us at all. The formula $\forall x F(x)$ would not have got among the antecedent formulas of the conclusion anyway, because there is a succedent formula $\forall x F(x)$ standing on the left of the main premise.

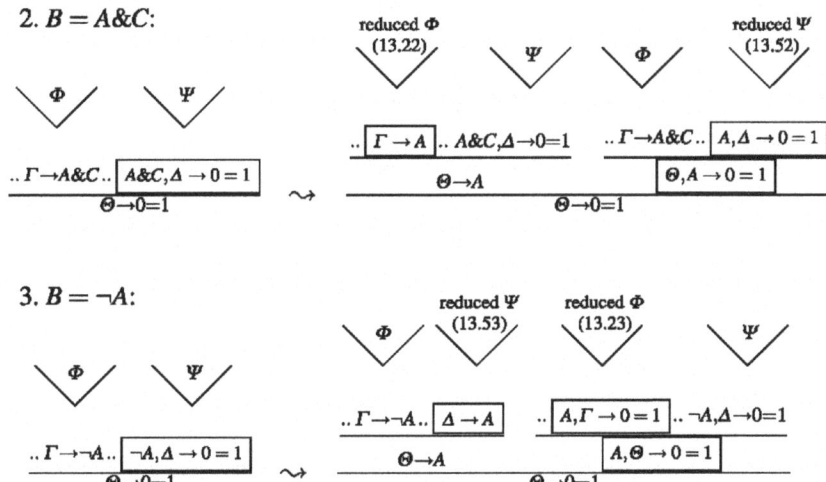

We would like to present an illustrative example to justify the freedom of choice in reduction steps 13.21 and 13.22. Imagine that we have this old derivation whose last inference rule is $\forall$-elimination:

$$\frac{\Gamma \to \forall x(x+0=0)}{\Gamma \to \bar{3}+0=0}$$

This derivation is translated into the following new derivation:

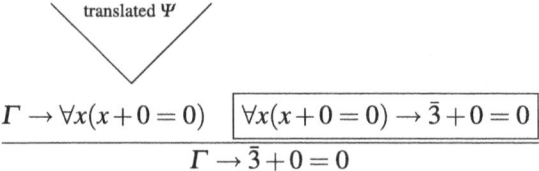

$$\frac{\Gamma \to \forall x(x+0=0) \quad \boxed{\forall x(x+0=0) \to \bar{3}+0=0}}{\Gamma \to \bar{3}+0=0}$$

The last inference rule is the chain rule and the main premise is a groundsequent. We know that this groundsequent is reduced to endform in the same way as $\forall x(x+0=0) \to \forall x(x+0=0)$ under the condition that the first reduction step of $\forall x(x+0=0) \to \forall x(x+0=0)$ consists in replacing x from the succedent by the numeral $\bar{3}$. This leads to $\forall x(x+0=0) \to \bar{3}+0=0$. Since there is a false atomic sentence in the succedent, we switch to steps 13.5 that modify the antecedent. We have to make the same adjustments in the same order as while modifying the succedent. So, we obtain $\bar{3}+0=0 \to \bar{3}+0=0$ and we are in endform.

Let $\Gamma \to \forall x(x+0=0)$ have a derivation whose reduction changes its endsequent according to 13.2. It is clear that the derivation from our example needs to be reduced according to e3: The main premise has a false atomic sentence in its succedent and the affected antecedent formula $\forall x(x+0=0)$ did not get among the antecedent formulas of the endsequent. The reduced derivation is:

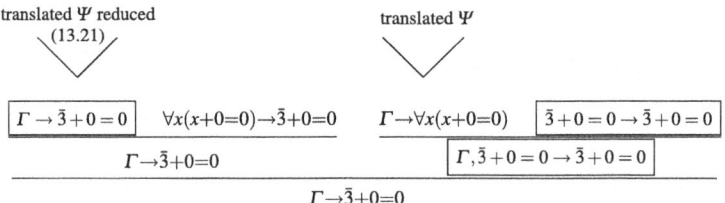

We see that we have no option but to carry out the reduction of the derivation of $\Gamma \to \forall x(x+0=0)$ in the way that the reduced sequent occurs in the form $\Gamma \to \bar{3}+0=0$. Another natural question is whether we are able to organize the reduction in this way. The premise $\Gamma \to \forall x(x+0=0)$ must have been derived by one of these rules:

- It is an initial sequent.
- ∀-introduction
- Chain rule that needs reduction according to e1.

As far as the first and the second item are concerned, it is clear that we are allowed to choose any numeral that we need. The reduction according to e1 tells us to choose the same numeral as for the main premise. Fortunately, the main premise was also reduced according to 13.2, so this is not a big restriction. The main premise must have been derived by one of the rules written above, too. Once we reach the 'premise of the premise' that is an initial sequent or was derived by ∀-introduction, we choose the numeral we require. This one is then inherited downwards until we are back in the endsequent.

Let us return to the analysis of case e3. We are going to compare the ordinal numbers of the original and the reduced derivation.

4.2 Lowering the Ordinal Numbers After Reduction Steps for Derivations

Lemma e3 *Reduction e3 does not change the numerus. The rank of the mantissa increases exactly by 2. The ordinal number of the reduced derivation is smaller than the one of the original derivation because the new mantissa is smaller than the old one.*

Proof Assume that the derivation had this form before we have reduced it:

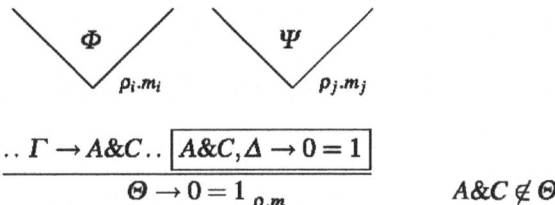

Assume that the maximal rank of all parental mantissae is ν. Hence, the rank of m is $\nu + 2$.

We start with an analysis of the two supporting chain rules from the reduced derivation, the left one and the right one:

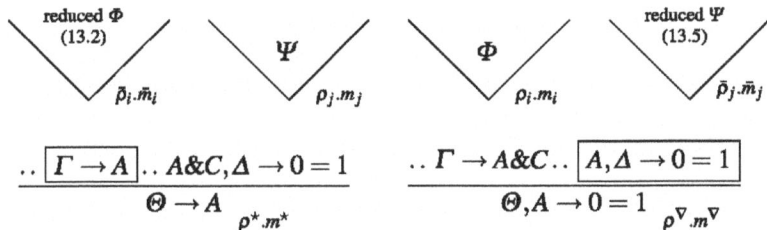

The premise $\Gamma \to A\&C$ was modified according to step 13.2 while reducing its derivation. Hence, one of these possibilities must be valid:

- It is an initial sequent.
- It is derived by the negation rule.
- It is derived by $\forall$-introduction.
- It is derived by the chain rule that needs reduction according to e1.

When we consider the reduction steps for derivations whose last inference rule is one of the mentioned above, we obtain:

- $\bar{\rho}_i = \rho_i$
- $\bar{m}_i < m_i$
- The mantissae $\bar{m}_i$ and m_i have the same rank.

The premise $A\&C, \Delta \to 0 = 1$ was modified according to step 13.5 while reducing its derivation. Hence, one of these possibilities must be valid:

- It is an initial sequent.
- It is derived by the chain rule that needs reduction according to e2.

We obtain the following result from the definition of reduction steps for initial sequents and from Lemma e2:

- $\bar{\rho}_j = \rho_j$
- $\bar{m}_j < m_j$
- The mantissae $\bar{m}_j$ and m_j have the same rank.

Now, we make a step further and compare the ordinal number $\rho.m$ of the original derivation with the numbers $\rho^\star.m^\star$ and $\rho^\nabla.m^\nabla$ which belong to the supporting chain rules, respectively:

Numerus $\rho^\star$: The maximal rank of the parental mantissae in the left supporting chain rule is the same as in the original derivation. It means that it is ν. The previous analysis gave us that m_i and $\bar{m}_i$ have the same rank and there are no other parental mantissae that have been changed. Since $\rho_i = \bar{\rho}_i$, the maximal excess of the parental numbers did not change, either. The only condition that could have become different, compared to the original derivation, is the maximal number of the logical operations in the succedent of a premise that stands on the left of the main premise. This number can be smaller in the examined supporting chain rule because there are fewer premises on the left of the main premise than in the original derivation. Thus, $\rho^\star \leq \rho$.

Mantissa $m^\star$: The mantissae m and $m^\star$ have the same rank $\nu + 2$. The mantissa $m^\star$ is built of the same sub-mantissae as m with the exception of $\bar{m}_i$ which is smaller than m_i. That is why we obtain $m^\star < m$.

Numerus ρ^∇: We have $\rho^\nabla = \rho$. The explanation for this is very similar to the one for $\rho^\star$. We even have that the maximal number of the logical operations in the succedent of a premise that stands on the left of the main premise is the same as in the original derivation.

Mantissa m^∇: It holds $m^\nabla < m$ and their ranks are the same: $\nu + 2$. For an explanation, see case $m^\star$.

Let us include the third chain rule of the reduced derivation:

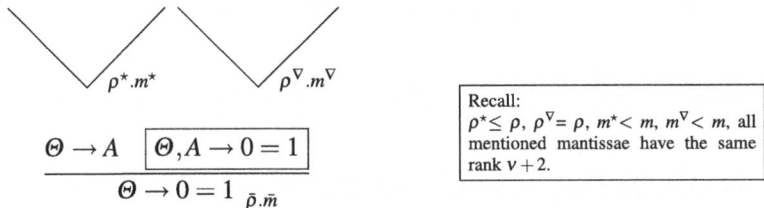

Recall:
$\rho^\star \leq \rho$, $\rho^\nabla = \rho$, $m^\star < m$, $m^\nabla < m$, all mentioned mantissae have the same rank $\nu + 2$.

We need to show that $\bar{\rho}.\bar{m} < \rho.m$. As usual, we are going to compare the numeri and the mantissae separately. The easier part is the examination of the relation between the mantissae m and $\bar{m}$. Let us launch into it.

The mantissa $\bar{m}$ starts with the greater parental mantissa. Assume it is $m^\star$.[2] The mantissa $m^\star$ as a part of $\bar{m}$ is followed by a sequence of 0's that is $\nu + 3$, in general $\nu + 4$, in length. We know that $m^\star$ is smaller than m. So, (1) the first digit that these two differ in is smaller in $m^\star$ or (2) $m^\star$ is an initial part of m. This implies that $\bar{m} < m$: As mentioned, $\bar{m}$ starts with $m^\star$. So, $\bar{m}$ and m either differ in a digit coming

---

[2] The analysis would be the same even if we had chosen m^∇.

4.2 Lowering the Ordinal Numbers After Reduction Steps for Derivations

from $m^\star$ or $m^\star$ is an initial part of both $\bar{m}$ and m. The point is that whereas $m^\star$ as a part of $\bar{m}$ is followed by a sequence of at least $\nu + 3$ zeros, there can be maximally $\nu + 2$ zeros in a row in m and this one does not end with 0 for sure:

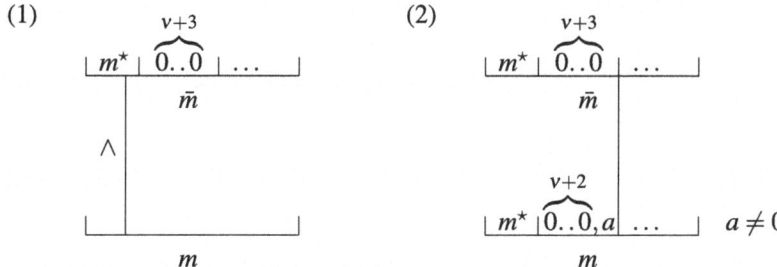

Note that although the rank of the mantissa m is $\nu+2$, the rank of $\bar{m}$ is two greater, hence $\nu + 4$.

Our last and most important task in case e3 is to compare ρ and $\bar{\rho}$. Since we have calculated the ranks of m and $\bar{m}$, we know:

- $\rho \geq \nu + 2$
- $\bar{\rho} \geq \nu + 4$

The desired result is $\bar{\rho} = \rho$, but it does not develop this way by now. Fortunately, there are two other conditions to satisfy. Let us consider the condition for the maximal number of the logical operations in the succedent of a premise that stands on the left of the main premise:

- $\rho \geq \nu + 2 + 2 \cdot |A \& C| = \nu + 2 + 2(x + i) = \nu + 2 + 2x + 2i$ where $|A| = x \geq 0$ and $i \geq 1$.
- $\bar{\rho} \geq \nu + 4 + 2 \cdot |A| = \nu + 4 + 2x$

We know $\rho = \nu + 2 + \text{excess}_{\rho.m}$ and we see that $\text{excess}_{\rho.m} \geq 2x + 2i \geq 2$. Since the mantissae m and m^∇ have the same rank and $\rho = \rho^\nabla$, we have $\text{excess}_{\rho.m} = \text{excess}_{\rho^\nabla.m^\nabla}$.

The definition of the numerus implies:

$$\bar{\rho} \geq \nu + 4 + \text{excess}_{\rho^\nabla.m^\nabla} - 2 = \nu + 2 + \text{excess}_{\rho.m}.$$

We used $\text{excess}_{\rho^\nabla.m^\nabla}$ because it is the maximal excess of the parental numbers. It is clear because $\rho^\star \leq \rho = \rho^\nabla$ and the ranks of $m^\star$ and m^∇ are the same.

We discovered that the numerus $\bar{\rho}$ is the *smallest* natural number that satisfies:

- $\bar{\rho} \geq \nu + 4 + 2x$ where $x \geq 0$.
- $\bar{\rho} \geq \nu + 2 + \text{excess}_{\rho.m}$ where $\text{excess}_{\rho.m} \geq 2$.

If we had

$$\nu + 4 + 2x \leq \underbrace{\nu + 2 + \text{excess}_{\rho.m}}_{\rho}$$

we would obtain the desired result $\bar{\rho} = \rho$.

What would happen if the contrary held? Assume $\nu + 2 + \text{excess}_{\rho.m} < \nu + 4 + 2x$:

$$\nu + 4 + 2x - (\nu + 2 + \text{excess}_{\rho.m}) > 0$$
$$\nu + 4 + 2x - \nu - 2 - \text{excess}_{\rho.m} > 0$$
$$2 + 2x - \text{excess}_{\rho.m} > 0$$
$$2 + 2x > \text{excess}_{\rho.m}$$

We already know that $\text{excess}_{\rho.m} \geq 2x + 2i$ where $i \geq 1$. Clearly, we obtained a contradiction. □

(**e4**): The endsequent $\Sigma \to 0 = 1$ is derived by chain rule ϑ_1. The main premise $\Theta \to 0 = 1$ is not in endform, so there exists a reduction step for its derivation. Assume the main premise remains unchanged after the reduction step was applied to its derivation and the last inference rule used in its derivation is chain rule ϑ that needs the reduction treated in item e3. The reduction of the whole derivation looks like this:

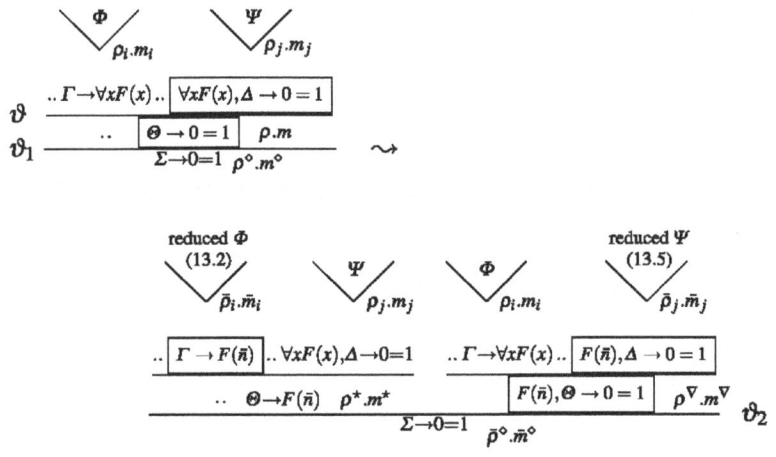

Picture e4

There are two new premises $\Theta \to F(\bar{n})$ and $F(\bar{n}), \Theta \to 0 = 1$ instead of $\Theta \to 0 = 1$ in the reduced derivation. They are results of the supporting chain rules that are similar to those from reduction e3. The difference is that we omit the third chain rule described in e3 and just place the results of the supporting chain rules at the position where $\Theta \to 0 = 1$ was before. The endsequent $\Sigma \to 0 = 1$ remains unchanged.

Lemma e4 *Reduction e4 does not change the numerus and the rank of the mantissa. The ordinal number of the reduced derivation is smaller than the one of the original derivation because the new mantissa is smaller than the old one.*

Proof We are going to work with Picture e4. It is obvious:

- Lemma e1 gives us $\rho_i = \bar{\rho}_i$ and $\bar{m}_i < m_i$. The mantissae $\bar{m}_i$ and m_i have the same rank.

4.2 Lowering the Ordinal Numbers After Reduction Steps for Derivations

- Lemma e2 gives us $\rho_j = \bar{\rho}_j$ and $\bar{m}_j < m_j$. The mantissae $\bar{m}_j$ and m_j have the same rank.
- Lemma e3 gives us: $\rho^\star \leq \rho, m^\star < m, \rho^\nabla = \rho, m^\nabla < m$. The mantissae $m^\star, m^\nabla$ and m have the same rank.

We wish to compare the ordinal numbers $\rho^\diamond.m^\diamond$ and $\bar{\rho}^\diamond.\bar{m}^\diamond$:

- We know that the sequent $\Theta \to 0 = 1$ is derived by chain rule ϑ. Assume that the maximal rank of all mantissae that are parental to this chain rule is ν. Since the ranks of m_i and $\bar{m}_i$ as well as the ranks of m_j and $\bar{m}_j$ are the same, we know that the maximal rank of the mantissae that are parental to the left and to the right supporting chain rule, respectively, is ν, too.

We can deduce that $m^\diamond$ and $\bar{m}^\diamond$ have the same rank $\nu + 2 + i$ where $i \geq 2$: The derivations of the premises $\Theta \to 0 = 1, \Theta \to F(\bar{n})$ and $F(\bar{n}), \Theta \to 0 = 1$ have ordinal numbers whose mantissae $m, m^\star$ and m^∇ have the ranks $\nu + 2$. The mentioned premises take part in chain rules ϑ_1 and ϑ_2. There are further premises with them whose derivations have some ordinal numbers, too. These numbers affect the maximal rank of the parental mantissae that is used in calculations for chain rules ϑ_1 and ϑ_2, respectively. Anyway, these ordinal numbers are the same in both rules. It follows that the rank of $m^\diamond$ and $\bar{m}^\diamond$ is $\nu + 2 + j + 2$ where $j \geq 0$. After a short modification we obtain $\nu + 2 + i$ where $i \geq 2$.

We have $m^\star < m$ and $m^\nabla < m$. Hence, the mantissa m from the original derivation was replaced by two smaller mantissae. The result is a global decrease, thus $\bar{m}^\diamond < m^\diamond$.

- We have $\rho^\diamond \geq (\nu + 2 + i)$ and $\bar{\rho}^\diamond \geq (\nu + 2 + i)$ where $i \geq 2$. Now, we are going to examine $\text{excess}_{\rho^\diamond.m^\diamond}$ and $\text{excess}_{\bar{\rho}^\diamond.\bar{m}^\diamond}$. These two values depend on the maximal excess of the parental numbers and on the maximal number of the logical operations in the succedent of a premise that stands on the left of the main premise. The maximal excess of the parental numbers is the same in both derivations, concretely, it is either $\rho - (\nu + 2) = \rho^\nabla - (\nu + 2)$ or it belongs to the ordinal number of a premise that is not drawn and that is represented by the dots (Picture e4).

When we try to find out the maximal number of logical operations in the succedent of a premise that stands on the left of the main premise, there are again two possibilities. The maximal number belongs either to a premise that is not drawn and we obtain $\rho^\diamond = \bar{\rho}^\diamond$ or we discover that the succedent formula in the reduced derivation we are looking for is $F(\bar{n})$. It follows that the examined number of logical operations in the original derivation is equal to or less than $|F(\bar{n})|$. Note that if it were not, it would be definitely caused by a formula that is not drawn. The same formula is in the reduced derivation too, and this would contradict the fact that $F(\bar{n})$ contains the most logical operations among all succedent formulas that stand on the left of the main premise. The conditions for the maximal number of logical operations for this case are:

- $\rho^\diamond - (\nu + 2 + i) = \text{excess}_{\rho^\diamond.m^\diamond} \geq 2 \cdot a$
- $\bar{\rho}^\diamond - (\nu + 2 + i) = \text{excess}_{\bar{\rho}^\diamond.\bar{m}^\diamond} \geq 2 \cdot |F(\bar{n})|$

Here $i \geq 2$ and $a \leq |F(\bar{n})|$. We are in a situation similar to item e3. It does not develop the way we need by now, but that does not matter: We are able to show that the condition for the logical operations has no influence in this case because we have:

$$\underbrace{(\text{excess}_{\rho.m} - 2)}_{\text{excess}_{\rho^\diamond.m^\diamond} \text{ and excess}_{\bar{\rho}^\diamond.\dot{m}^\diamond} \text{ have at least this value}} \geq 2 \cdot |F(\bar{n})|$$

Let us calculate:
$\text{excess}_{\rho.m} - 2 \geq 2 \cdot |\forall x F(x)| - 2 = 2(y+1) - 2 = 2y$ where $y = |F(x)|$

We showed the desired result. Since $\text{excess}_{\rho.m} = \text{excess}_{\rho^\nabla.m^\nabla}$ and a numerus is the smallest natural number satisfying the three conditions formulated in Sect. 4.1, it is clear that the maximal excess of the parental numbers has the crucial role while calculating the numeri $\rho^\diamond$ and $\bar{\rho}^\diamond$ and the maximal number of the logical operations carries no weight. We obtain $\bar{\rho}^\diamond = \rho^\diamond$. □

(**e5**): The endsequent is derived by the chain rule. The main premise is not in endform, so there exists a reduction step for its derivation. Assume that the main premise is transformed according to reduction step 13.5 while reducing its derivation, and that the affected formula B is not among the antecedent formulas of the endsequent. It was deleted because it is the succedent formula of a premise that stands on the left of the main premise:

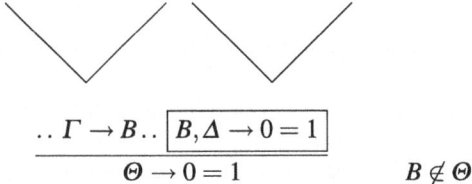

Since B was used in reduction step 13.5, it is not atomic. So, by the induction hypothesis, a reduction step for the derivation of the premise $\Gamma \to B$ is defined. Assume that the premise $\Gamma \to B$ remains unchanged after the reduction step was applied to its derivation, and that the last inference rule used in its derivation is Chain rule that needs reduction treated in item e3. This is the described case:

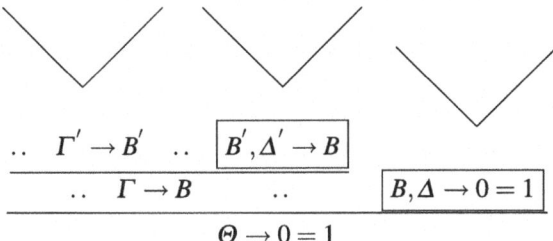

We said that the derivation of the premise $\Gamma \to B$ needs reduction according to e3. It follows that B must have the form $0 = 1$. This contradicts the fact that B is not atomic, thus, this case is impossible.

The following cases are very similar to cases e4 and e5. The only difference is that the sequent which remains unchanged after its derivation was reduced had not been

4.2 Lowering the Ordinal Numbers After Reduction Steps for Derivations

derived by the chain rule that needs reduction according to e3. It had simply been derived by another rule that, when used as the last inference rule in a derivation, does not change the endsequent after the reduction. The similarity between cases e4, e5, e6, e7 is probably the reason why they are studied as one case by Gentzen. We shall also provide one proof concerning the decrease of the ordinal numbers for cases e6, e7 and this will be similar to those for Lemmas e1 and e2.

(**e6**): The endsequent is derived by the chain rule. The main premise is not in endform, so there exists a reduction step for its derivation. Assume that the main premise remains unchanged after the reduction step was applied to its derivation, and that the last inference rule used in its derivation *is not* a chain rule that needs the reduction treated in item e3.

The reduction is very simple. We take the reduced derivation of the main premise. The endsequent of this derivation is the main premise itself. This is an advantage because we do not have to adjust anything else. The chain rule is correct and the endsequent of the original as well as of the reduced derivation is the same:

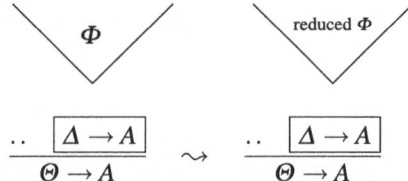

(**e7**): The endsequent is derived by the chain rule. The main premise is not in endform, so there exists a reduction step for its derivation. Assume that the main premise is transformed according to reduction step 13.5 while reducing its derivation, and that the affected formula B is not among the antecedent formulas of the endsequent. It was deleted because it is the succedent formula of a premise that stands on the left of the main premise:

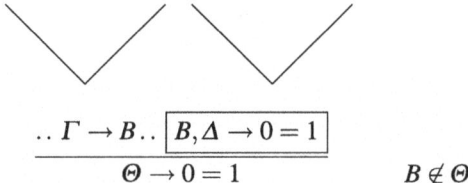

Since B was used in reduction step 13.5, it is not atomic. So, by the induction hypothesis, a reduction step for the derivation of the premise $\Gamma \to B$ is defined. Assume that the premise $\Gamma \to B$ remains unchanged after the reduction step was applied to its derivation, and that the last inference rule used in its derivation *is not* a Chain rule that needs reduction treated in item e3. The reduction looks like this:

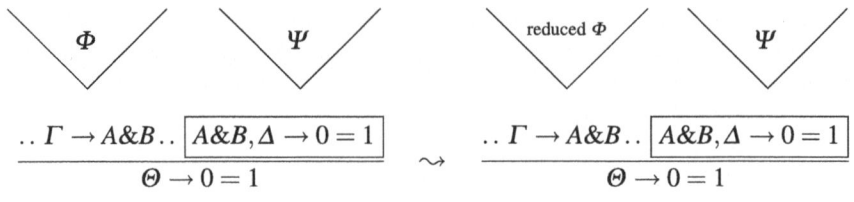

Picture e7

The reduction step was carried out on the derivation of the unchanged premise $\Gamma \to A\&B$. The reduced derivation of the main premise was not used. The chain rule remains correct and the endsequent of the original as well as of the reduced derivation is the same.

We are going to show that both reductions e6 and e7 lower the ordinal numbers.

Lemma e6–e7 *The numerus does not increase after reductions e6 and e7. The rank of the mantissa remains the same. The new mantissa is smaller than the old one. It follows that the ordinal number of the reduced derivation is smaller than the one of the original derivation.*

Proof Let us take an inference step in our derivation that uses a chain rule which needs to be reduced according to e6 or e7, such that there is *no* inference step above the chosen step that has to be reduced according to e6 or e7, too.

Assume first that the inference step we found needs reduction e6. Thus, its main premise must have been derived by one of these rules:

- induction rule
- Chain rule that needs reduction according to e4.

The main premise could not have been derived by a chain rule that needs reduction e3, because it is forbidden in the description of the case. Similarly, it could not have been derived by a chain rule that needs reduction e6 or e7. That would contradict the initial choice. Since reductions of the derivations whose last inference rule is another one than mentioned above modify the endsequent, they are not possible.

Let us assign ordinal numbers to the derivations:

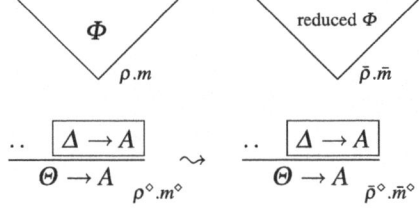

The following holds for the ordinal numbers of the derivations of the main premise $\Delta \to A$:

- If it is derived by the induction rule, then $\bar{\rho} \leq \rho$ and $\bar{m} < m$. The mantissae $\bar{m}$ and m have the same rank.

4.2 Lowering the Ordinal Numbers After Reduction Steps for Derivations

- If it is derived by the chain rule that needs reduction e4, then $\bar{\rho} = \rho$ and $\bar{m} < m$. The mantissae $\bar{m}$ and m have the same rank.

Let us compare numbers $\rho^\diamond.m^\diamond$ and $\bar{\rho}^\diamond.\bar{m}^\diamond$ of the whole derivations:

- The mantissae $m^\diamond$ and $\bar{m}^\diamond$ have the same rank.
- $\bar{m}^\diamond < m^\diamond$
- $\bar{\rho}^\diamond \leq \rho^\diamond$

The second possibility is that the chain rule we found needs to be reduced according to e7. Hence, the premise $\Gamma \to A\&B$ (see Picture e7) must have been derived by one of these rules:

- induction rule
- Chain rule that needs reduction according to e4.

These are actually the same as in case e6.

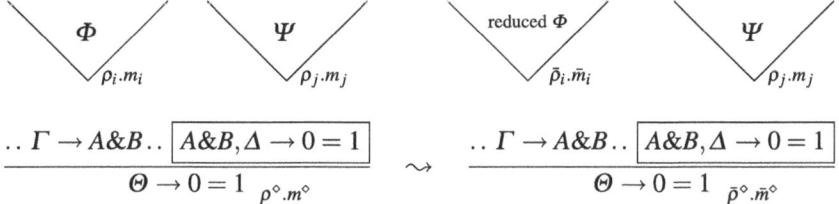

The usual explanation gives us that $\bar{\rho}^\diamond \leq \rho^\diamond$ and $\bar{m}^\diamond < m^\diamond$. The mantissae $m^\diamond$ and $\bar{m}^\diamond$ have the same rank.

Now, the induction hypothesis follows: Assume the ordinal number of a derivation whose endsequent is derived by the chain rule that needs reduction e6 or e7 gets smaller after the reduction. The reduction makes it smaller because the new mantissa is smaller than the old one. The numerus does not get greater, it can be even smaller, and the rank remains the same. We need to show that this holds for the next sequent derived by the chain rule which also needs reduction e6 or e7. The analysis would look like the analysis above that led to the induction hypothesis. □

The analysis of all possible cases is completed. We have proved that all reduction steps for derivations decrease the ordinal number of the reduced derivation. Since the ordinal numbers are well-ordered, we have to reach a derivation that cannot be reduced anymore. Thus, its endsequent is in endform. If it were not, we would be able to reduce it again. The derivation of the sequent $\to 0 = 1$ cannot be reduced in the way that modifies the form of the endsequent because a reduction step for sequents applicable to $\to 0 = 1$ does not exist. At the same time, we see that $\to 0 = 1$ is not in endform. We can conclude that we should be able to reduce the derivation of the sequent $\to 0 = 1$ infinitely many times and every reduction step is expected to make the ordinal number smaller. This contradicts the well-ordering of the ordinal numbers and we obtain that it is impossible to derive the sequent $\to 0 = 1$ in PA.

Appendix A
Removal of Logical Operations $\vee$, $\supset$ and $\exists$ from a Derivation

We replace the logical operations $\vee$, $\supset$ and $\exists$ in an arbitrary derivation in PA by the following equivalent expressions:

$$(A \vee B) \leftrightarrow \neg(\neg A \& \neg B)$$
$$\exists x F(x) \leftrightarrow \neg \forall x \neg F(x)$$
$$(A \supset B) \leftrightarrow \neg(A \& \neg B)$$

Now it is turn to make the derivation correct again:

- Each logical initial sequent $D \to D$ remains unchanged or turns into a different one, but still a logical initial sequent.
- The modification of a mathematical initial sequent $\to C$, where C is an equality axiom or a Robinson arithmetic axiom, does not spoil it because we deal with equivalent adjustments in classical predicate logic. The simplest solution would be not to use the operations $\vee$, $\supset$ and $\exists$ in our axioms at all.
- The induction rule and the structural rules remain correct.
- The logical rules for $\&$, $\neg$ and $\forall$ remain correct.
- The logical rules for $\vee$, $\supset$ and $\exists$ are not correct anymore and it is necessary to fix them.

In this section, a formula A^* stands for the equivalent modification of A as described earlier. A sequence of formulas with a star, for example Γ^*, stands for a sequence of formulas where every formula has a star.

Rule $\vee$-*introduction* had this form: $\frac{\Gamma \to A}{\Gamma \to A \vee B}$. We have after the modification: $\frac{\Gamma^* \to A^*}{\Gamma^* \to \neg(\neg A^* \& \neg B^*)}$. We correct it by replacing it with the following derivation:

$$\frac{Wk \dfrac{\Gamma^* \to A^*}{\neg A^* \& \neg B^*, \Gamma^* \to A^*} \qquad \dfrac{\neg A^* \& \neg B^* \to \neg A^* \& \neg B^*}{\neg A^* \& \neg B^* \to \neg A^*} \& E}{\Gamma^* \to \neg(\neg A^* \& \neg B^*)} \neg I$$

72 Appendix A: Removal of Logical Operations ∨, ⊃ and ∃ from a Derivation

Rule ∨-*elimination* had this form: $\frac{\Gamma \to A \vee B \quad A, \Delta \to C \quad B, \Theta \to C}{\Gamma, \Delta, \Theta \to C}$. We have now: $\frac{\Gamma^* \to \neg(\neg A^* \& \neg B^*) \quad A^*, \Delta^* \to C^* \quad B^*, \Theta^* \to C^*}{\Gamma^*, \Delta^*, \Theta^* \to C^*}$. We correct it by replacing it with the following derivation:

$$\neg I, Ex \frac{\&I, Ct \frac{\neg I \frac{Wk \frac{\neg C^* \to \neg C^*}{A^*, \neg C^* \to \neg C^*} \quad A^*, \Delta^* \to C^*}{\neg C^*, \Delta^* \to \neg A^*} \quad \neg I \frac{\frac{\neg C^* \to \neg C^*}{B^*, \neg C^* \to \neg C^*} \quad B^*, \Theta^* \to C^*}{\neg C^*, \Theta^* \to \neg B^*}}{\neg C^*, \Delta^*, \Theta^* \to \neg A^* \& \neg B^*} \quad Wk \frac{\Gamma^* \to \neg(\neg A^* \& \neg B^*)}{\neg C^*, \Gamma^* \to \neg(\neg A^* \& \neg B^*)}}{\neg E \frac{\Gamma^*, \Delta^*, \Theta^* \to \neg\neg C^*}{\Gamma^*, \Delta^*, \Theta^* \to C^*}}$$

Rule ∃-*introduction* had this form: $\frac{\Gamma \to F(t)}{\Gamma \to \exists x F(x)}$. It looks like this after the modification: $\frac{\Gamma^* \to F(t)^*}{\Gamma^* \to \neg \forall x \neg F(x)^*}$. This must be replaced with:

$$\frac{Wk \frac{\Gamma^* \to F(t)^*}{\forall x \neg F(x)^*, \Gamma^* \to F(t)^*} \quad \forall E \frac{\forall x \neg F(x)^* \to \forall x \neg F(x)^*}{\forall x \neg F(x)^* \to \neg F(t)^*}}{\Gamma^* \to \neg \forall x \neg F(x)^*} \neg I$$

Rule ∃-*elimination* had this form: $\frac{\Gamma \to \exists x F(x) \quad F(a), \Delta \to C}{\Gamma, \Delta \to C}$. We have now: $\frac{\Gamma^* \to \neg \forall x \neg F(x)^* \quad F(a)^*, \Delta^* \to C^*}{\Gamma^*, \Delta^* \to C^*}$. We use this derivation instead of the incorrect rule:

$$\neg I, Ex \frac{\forall I \frac{\neg I \frac{Wk \frac{\neg C^* \to \neg C^*}{F(a)^*, \neg C^* \to \neg C^*} \quad F(a)^*, \Delta^* \to C^*}{\neg C^*, \Delta^* \to \neg F(a)^*}}{\neg C^*, \Delta^* \to \forall x \neg F(x)^*} \quad Wk \frac{\Gamma^* \to \neg \forall x \neg F(x)^*}{\neg C^*, \Gamma^* \to \neg \forall x \neg F(x)^*}}{\neg E \frac{\Gamma^*, \Delta^* \to \neg \neg C^*}{\Gamma^*, \Delta^* \to C^*}}$$

There is a ∀-introduction used in the derivation. It requires the eigenvariable a not to occur in $\neg C^*, \Delta^*, \forall x \neg F(x)^*$. We used ∃-elimination before. So, as an assumption, we know that the eigenvariable a does not occur in $\Gamma, \Delta, C, \exists x F(x)$. It follows that the condition for the eigenvariable a imposed by ∀-introduction is satisfied.

Rule ⊃-*introduction* had this form: $\frac{A, \Gamma \to B}{\Gamma \to A \supset B}$. It looks like this after the modification: $\frac{A^*, \Gamma^* \to B^*}{\Gamma^* \to \neg(A^* \& \neg B^*)}$. We replace it with this derivation:

$$\neg I \frac{\neg I, Ex \frac{Wk \frac{\&E \frac{A^* \& \neg B^* \to A^* \& \neg B^*}{A^* \& \neg B^* \to \neg B^*}}{A^*, A^* \& \neg B^* \to \neg B^*} \quad A^*, \Gamma^* \to B^*}{\Gamma^*, A^* \& \neg B^* \to \neg A^*} \quad \&E \frac{A^* \& \neg B^* \to A^* \& \neg B^*}{A^* \& \neg B^* \to A^*}}{\Gamma^* \to \neg(A^* \& \neg B^*)}$$

Rule ⊃-*elimination* had the following form: $\frac{\Gamma \to A \quad \Delta \to A \supset B}{\Gamma, \Delta \to B}$. We obtain after the modification: $\frac{\Gamma^* \to A^* \quad \Delta^* \to \neg(A^* \& \neg B^*)}{\Gamma^*, \Delta^* \to B^*}$. The solution is:

Appendix A: Removal of Logical Operations $\vee$, $\supset$ and $\exists$ from a Derivation

$$\neg I \dfrac{\&I \dfrac{\Gamma^* \to A^* \quad \neg B^* \to \neg B^*}{\Gamma^*, \neg B^* \to A^*\&\neg B^*} \quad \dfrac{\Delta^* \to \neg(A^*\&\neg B^*)}{\neg B^*, \Delta^* \to \neg(A^*\&\neg B^*)} Wk}{\neg E \dfrac{\Gamma^*, \Delta^* \to \neg\neg B^*}{\Gamma^*, \Delta^* \to B^*}}$$

Finally, our derivation is correct and does not contain the operations $\vee$, $\supset$ and $\exists$.

Index

Symbols
| A |, 14
LK, 5
NJ, 3
NK, 3
NLK, 11, 14, 15
    logical rules, 14
    structural rules, 12
Überschuß, 42

A
Ackermann, 3, 5
Affected formula, 20, 23
Algorithm
    initial sequents to endform, 21
    ordinal numbers to Cantor normal form, 35
        examples, 39
An sich-concept, 6
Axioms
    equality axioms, 12
    Robinson arithmetic axioms, 12

B
Bernays, 2–4
Beseitigung der doppelten Verneinung, 7, 13
Brouwer-Heyting-Kolmogorov interpretation, 3, 6

C
Cantor normal form, 5, 8
    definition, 35
Chain rule, 24, 27
    definition, 25

    example, 25
    explanation, 24
Conclusion
    definition, 12
Consistency proof
    absolute, 2
    relative, 2
Contraction, 27
Cut, 14
Cut elimination theorem, 3

D
Derivation
    definition, 14
    difference between proof and derivation, 14
    new, 15
        definition, 24
    old, 15
        definition, 14
    translate *old* into *new*, 26

E
Eigenvariable, 13, 15
Endform, 15–19
    definition, 14
Endsequent, 14
Endsequenz, 14
Excess, 42
Exchange, 27

F
Fan theorem, 4
Finitistic

interpretation of logical operations, 6, 7
   methods, 2, 7
Formalism, 1
Formulas
   affected formula
      definition, 20
   antecedent formulas, 11
   succedent formula, 11
Freedom of choice, 16, 17, 19
   example, 59
Frege, 2

G
Gödel, 2–4, 6, 7
Gödel-Gentzen translation, 3
Gentzen's thesis, 2–6, 11
Groundsequents, 15, 23, 24

H
Hauptfall, 58
Hauptsatz, 3
Herleitung, 14
Hertz, 2
   systems, 4
Heyting, 3
Hilbert, 1, 3
Hilbert's program, 1
Hinterformel, 11
How to assign ordinal numbers
   chain rule, 43
   induction rule, 44
   initial sequent, 42
   ∀-introduction, 42
   negation rule, 42

I
Induction
   bar induction, 4
   rule, 3, 4, 13
   transfinite up to ε_0, 4, 5, 7, 18
Infinity
   actual, 1, 2, 6
   potential, 6
Intuitionism, 1, 5, 8

K
Kalmar, 5
Kettenschluß, 24
Kneser, 2, 4

L
Logicism, 2
Lowering of ordinal numbers
   chain rule, 50
      e1, 53
      e2, 55
      e3, 58
      e4, 64
      e5, 66
      e6, 67
      e7, 67
      Hauptfall, 58
      preparation step, 50
   induction rule, 47
   initial sequents, 46
   ∀-introduction, 46
   negation rule, 47

M
Main premise, 24
   definition, 25
Mantissa, 30
   parental, 42
Mantisse, 30

N
Naive set theory, 1, 6
Natural deduction, 2–4, 7
Natural deduction in sequent calculus style, 5, 6, 11
Negation rule, 24
Normalization, 3, 4, 7
Numerus, 30
   parental, 42

O
Ordinal analysis, 5
Ordinal numbers
   Cantor normal form, 35
   definition, 30
   examples, 30
   excess, 42
   general condition, 41
   mantissa, 30
   numerus, 30
   ordinal numbers to Cantor normal form, 35
      examples, 39
   parental mantissa, 42
   parental number, 42
   parental numerus, 42
   rank, 41

relationship with standard notation, 32
systems σ, 31
    example of use, 33
    ordering of systems σ, 31
    ordering within a single system σ, 32

P
Parental mantissa, 42
Parental number, 42
Parental numerus, 42
Premise
    definition, 12
    main premise, 24
        definition, 25
Proof
    difference between proof and derivation, 14

R
Rank, 41
Reduction
    groundsequents to endform, 22
    logical initial sequents to endform, 21
    mathematical initial sequents to endform, 21
    procedure, 4, 5, 7
    reduction steps for derivations, 16, 18
    reduction steps for sequents, 16, 17
        steps 13.2, 17, 19
        steps 13.5, 17–19
Reduziervorschrift, 4
Renaming of bound variables, 27
Rule
    chain rule, 15, 16, 18, 24, 27
        definition, 25
        example, 25
        explanation, 24
    cut, 14, 16
    induction rule, 13, 15, 16, 24
    logical rules, 14
    negation rule, 24

rules of inference, 12
structural rules, 12, 27
Russell, 2

S
Schütte, 4
Sequent calculus, 3–6
Sequents
    definition, 11
    logical initial, 12
    mathematical initial, 12
    new logical initial, 24
    new mathematical initial, 24
Strukturänderungen, 12
Subformula property, 3, 4
Systems σ, 31
    example of use, 33

T
Takeuti, 5

V
Verdünnung, 13
Vertauschung, 13
Von Neumann, 4
Vorderformeln, 11

W
Wahlfreiheit, 16, 19
    example, 59
Weakening, 27
Weyl, 4
Widerlegung, 13

Z
Zusammenziehung, 13

The manufacturer's authorised representative in the EU is Springer Nature Customer Service Centre GmbH, Europaplatz 3, 69115 Heidelberg, Germany. If you have any concerns regarding our products, please contact ProductSafety@springernature.com

Printed and bound by CPI Group (UK) Ltd, Croydon, CR0 4YY

23/03/2026

02076395-0015